Electric Arc Furnace Steelmaking

Electric Arc Furnace Steelmaking

Guest Editors

Thomas Echterhof
Ville-Valtteri Visuri

MDPI

Basel • Beijing • Wuhan • Barcelona • Belgrade • Novi Sad • Cluj • Manchester

Guest Editors

Thomas Echterhof
Department for Industrial Furnaces and Heat Engineering
RWTH Aachen University
Aachen
Germany

Ville-Valtteri Visuri
Process Metallurgy Research Unit
University of Oulu
Oulu
Finland

Editorial Office

MDPI AG
Grosspeteranlage 5
4052 Basel, Switzerland

This is a reprint of the Special Issue, published open access by the journal *Metals* (ISSN 2075-4701), freely accessible at: https://www.mdpi.com/journal/metals/special_issues/QYB38P69T9.

For citation purposes, cite each article independently as indicated on the article page online and as indicated below:

Lastname, A.A.; Lastname, B.B. Article Title. *Journal Name* **Year**, *Volume Number*, Page Range.

ISBN 978-3-7258-6230-6 (Hbk)
ISBN 978-3-7258-6231-3 (PDF)
https://doi.org/10.3390/books978-3-7258-6231-3

Contents

About the Editors

Thomas Echterhof

Thomas Echterhof is Academic Director and Deputy Head of the Department for Industrial Furnaces and Heat Engineering at RWTH Aachen University. He leads the research group on mass and energy balances, with a special focus on the sustainability of electric arc furnace (EAF) steelmaking, including circular economy aspects. His main research interests include improving energy and resource efficiency and reducing the environmental impact of energy- and resource-intensive processes such as EAF steelmaking. He is also actively involved in teaching and delivers lectures on electric arc furnace technology at RWTH Aachen University. He received his Dipl.-Ing. and Dr.-Ing. degrees from RWTH Aachen University in 2005 and 2010, respectively. He has authored more than 70 peer-reviewed publications in journals and conference proceedings.

Ville-Valtteri Visuri

Ville-Valtteri Visuri is an Associate Professor at the University of Oulu, where he leads a research group focusing on the primary metallurgy of steelmaking. He also serves as Deputy Head of the Process Metallurgy Research Unit and as Chair of the Executive Board of the Centre for Advanced Steels Research (CASR). His research focuses on the chemical and physical fundamentals of steelmaking processes, particularly electric arc furnaces, converters, hot metal desulfurization, and hydrogen plasma smelting reduction, using mathematical modelling, computational thermodynamics, and high-temperature experiments. He holds the title of Docent in Process Metallurgy (2020), a D.Sc. (Tech.) in Process Engineering (2017), and M.Sc. (Tech.) and B.Sc. (Tech.) degrees in Industrial Engineering and Management (2011 and 2009), all from the University of Oulu. He has authored more than 80 peer-reviewed publications in journals and conference proceedings.

Preface

This Special Issue represents a follow-up to our earlier, methodologically focused Special Issue entitled "Modeling and Simulation of Metallurgical Processes in Ironmaking and Steelmaking", which was completed in 2022, along with a corresponding Special Issue reprint published in the same year. In this Special Issue, we have chosen a process-oriented focus on electric arc furnace (EAF) steelmaking. The Department of Industrial Furnaces and Heat Engineering at RWTH Aachen University and the Process Metallurgy Research Unit at the University of Oulu have collaborated in this area for over a decade, initially concentrating on the use of optical emission spectroscopy for process monitoring and later expanding into modelling of the EAF process. We wish to thank our current and former colleagues at the University of Oulu and RWTH Aachen University for their active collaboration. The strong institutional support of Professor Timo Fabritius (University of Oulu) and Professor Herbert Pfeifer (RWTH Aachen University) is gratefully acknowledged, as without it the numerous research visits between the two institutions would not have been possible. Since 2015, RWTH Aachen University has organized the European Academic Symposium on EAF Steelmaking (EASES) biannually. As a milestone in the collaboration between RWTH Aachen University and the University of Oulu, the 5th EASES, which was the first to be held outside Aachen, was organized in Oulu in June 2023.

Following peer review, eight manuscripts were selected for publication in this Special Issue. These papers address many of the current hot topics in EAF research, including the use of direct reduced iron and non-fossil carbon. The editorial of this Special Issue aims to place these papers in perspective in light of recent research published elsewhere. We hope that the results and conclusions presented in this Special Issue will benefit both the academic community and industry, sparking new ideas for further studies and contributing to advancements in steelmaking practice.

Thomas Echterhof and Ville-Valtteri Visuri
Guest Editors

metals

MDPI

Editorial

Electric Arc Furnace Steelmaking

Ville-Valtteri Visuri [1,*] and Thomas Echterhof [2]

1 Process Metallurgy Research Unit, University of Oulu, P.O. Box 4300, 90014 Oulu, Finland
2 Department for Industrial Furnaces and Heat Engineering, RWTH Aachen University, Kopernikusstraße 10, 52074 Aachen, Germany; echterhof@iob.rwth-aachen.de
* Correspondence: ville-valtteri.visuri@oulu.fi

1. Introduction

Research and development regarding the electric arc furnace (EAF) started as early as the late 19th century. In 1888, Paul Héroult patented an EAF featuring direct heating of the metal charge. It became the basis for an industrial prototype in 1899 and for industrial-scale steel production in 1907 [1]. However, the EAF only became a major steelmaking process after World War II, with large-scale EAF plants emerging from the 1950s onward [2]. Since then, improvements in transformer capacity have enabled an increase in EAF capacities—from 5–25 tons in first-generation (regular power (RP)) furnaces to as much as 400 tons in fourth-generation (super-ultra-high power (SUHP)) furnaces [1].

Today, the EAF is unquestionably one of the key unit processes in modern steelmaking and a cornerstone of the mini mill concept [3]. The EAF is flexible in terms of both raw materials and energy input: scrap and other iron-bearing materials, such as direct-reduced iron (DRI) and hot-briquetted iron (HBI), can be melted in any ratio, and the electrical energy input can be supplemented with additional chemical energy from burners and injectors to expedite melting [3,4]. The DRI or HBI can be charged in buckets or fed continuously [5].

In 2024, electric steelmaking accounted for approximately 29% of global crude steel production of 1885 Mt [6]. In many highly developed regions, such as North America and Europe, the share is significantly higher. Considering the transformation strategies of integrated steelworks to decarbonize production, together with increasing scrap availability in Asia, it is expected that the share of scrap-based EAF steelmaking will continue to rise in the coming decades.

The global DRI production in 2024 was only 144 Mt [6], but it is geographically concentrated in countries with access to inexpensive natural gas, which is used as the reductant [7]. In that year, the top three producers—India, Iran, and Saudi Arabia—accounted for approximately two-thirds of global DRI production [6].

In recent years, ambitious plans to produce CO_2-lean steel by melting or smelting DRI obtained through hydrogen direct reduction (H-DR), with or without the use of accompanying natural gas, have been announced. Launched in 2017, the Hydrogen Breakthrough Ironmaking Technology (HYBRIT) initiative represents a collaboration between the Swedish companies Vattenfall, LKAB, and SSAB aiming to study steelmaking based on the H-DR–EAF route, with the ambitious goal of reducing CO_2 emissions to 25 kg per ton of steel [8]. To demonstrate hydrogen direct reduction, a pilot plant with a capacity of 1 ton per hour was constructed in Luleå, Sweden, and it became operational in 2020 [8]. The HYBRIT concept has attracted interest from both the general public and steelmaking companies. In the Nordic countries, the publicly announced plans of Blastr Green Steel (Inkoo, Finland) [9], Stegra (Boden, Sweden) [10], and SSAB (Luleå, Sweden) [11] all follow the H-DR–EAF route. In the case of limited DRI availability, this route allows flexible use of DRI–scrap mixtures. Both the total energy demand and the amount of slag generated increase with higher DRI/scrap ratios [12]. The overall slag volumes are lower than in the case of the blast furnace (BF)–basic oxygen furnace (BOF) route [13]. However, the high basicity and the prevalent RO phase, which lacks hydraulic properties, complicate the direct use of H-DRI-based EAF slags as a supplementary cementitious material (SCM) to compensate for the decreasing production of granulated blast furnace slag [13].

While the H-DR–EAF route is well suited for melting high-quality DRI, the high gangue content of low-quality DRI results in large EAF slag volumes due to the basicity requirements ($B_2 = CaO/SiO_2 \approx 2$) of the foaming slag practice [14]. In combination with FeO contents above 20 wt%, this leads to a significant loss of iron yield [14]. Consequently, processing low-quality DRI is expected to be more economical in an electric smelting furnace (ESF), with the resulting hot metal subsequently refined in a BOF [14,15]. An ESF is operated with a slag basicity of $B_2 \approx 1$, which is close to that of BF slag and is therefore expected to allow its direct use as an SCM, as with BF slag [14]. A pilot-scale ESF plant was built in Pori, Finland, by Metso [16]. Furthermore, the German conglomerate ThyssenKrupp has announced plans to implement the H-DR–ESF–BOF route at its Duisburg site [17]. The strong interest in DRI-based steelmaking has broadened the scope of EAF research to include the processing of H-DRI and has established ESFs as a significant new research area in steelmaking.

Global stainless-steel production has continued its strong growth in the 21st century, reaching a new record of 62.6 Mt in 2024 [18]. Modern stainless-steel production is predominantly based on scrap-based EAF steelmaking, followed by argon–oxygen decarburization (AOD), known as the duplex route, or a combination of AOD and vacuum oxygen decarburization (VOD), known as the triplex route [19]. Aside from reducing its direct CO_2 emissions, the efforts to decarbonize EAFs in stainless steelmaking are largely linked to lowering the CO_2 footprint with respect to both the electricity and raw materials used. In this regard, the decarbonization of ferrochrome production is of particular importance, as ferrochrome provides the chromium required for alloying.

Arc furnaces have recently attracted a substantial amount of additional research attention due to the strong academic interest in hydrogen plasma-smelting reduction (HPSR) [20], which can be applied either as a one-step process for melting, reduction, and refining or alongside pre-reduction using hydrogen [21]. Much of the research interest in HPSR, such as arc stability [22], impurity removal [23], and refractory concepts [24], aligns with the approaches studied in the field of EAF research.

In view of the developments outlined above, the importance of EAFs as a research area remains as significant as ever. In this Special Issue of *Metals*, we present contributions pertaining to recent advances in EAF steelmaking. This editorial is intended to provide an introduction to and overview of the papers included in the Special Issue as well as selected highlights.

2. Overview of the Contributions

A total of eight peer-reviewed articles, comprising seven original articles and one review article, were published as part of this Special Issue (see Table 1). The papers cover many of the key aspects of the EAF process, ranging from raw materials via melting and slag foaming to impurity removal.

Table 1. Overview of the articles in this Special Issue.

			Focus					
No.	**Article**	**Article Type** [1]	**Raw Materials**	**Melting**	**Slag Foaming**	**Impurity Removal**	**Modelling and Simulation**	**System Level**
1	Azzaz et al. (Contribution 1)	O				X	X	
2	Dall'Osto et al. (Contribution 2)	O	X	X				
3	Reinicke et al. (Contribution 3)	O					X	
4	Dall'Osto et al. (Contribution 4)	O						X
5	Ugarte et al. (Contribution 5)	O				X	X	
6	Kieush et al. (Contribution 6)	O	X		X			
7	Pfeiffer et al. (Contribution 7)	O	X	X				
8	Kieush et al. (Contribution 8)	R	X					

[1] Article types: O = original, and R = review.

The papers feature authors affiliated with institutions and companies from six countries (see Figure 1a), with universities being the most common affiliation for both first and subsequent authors (see Figure 1b). Nevertheless, some authors are affiliated with technology suppliers, research institutes, or steel producers. In total, the papers list 35 authorships, representing 30 unique individuals, with an average of four authors per paper.

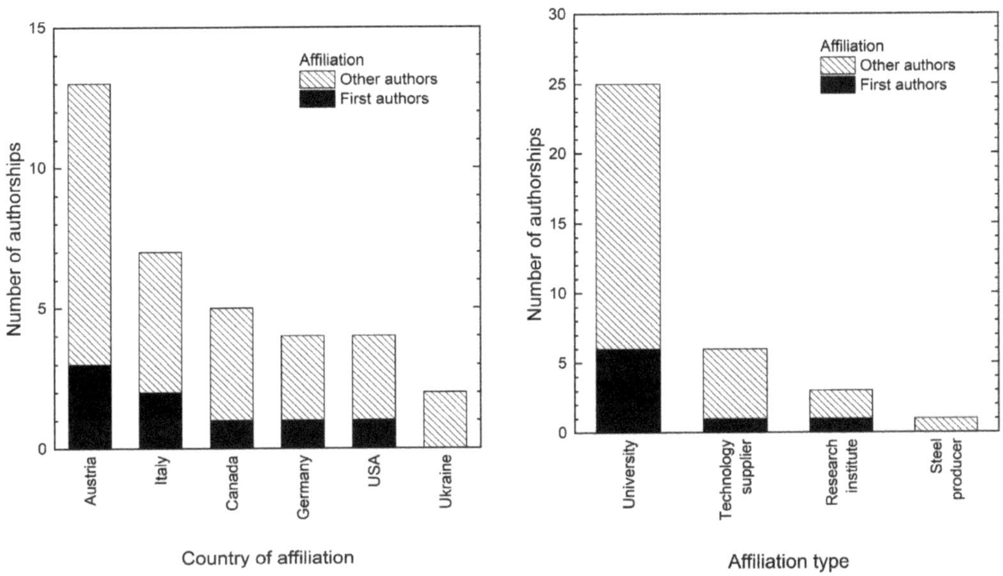

(**a**) (**b**)

Figure 1. Number of authorships by (**a**) country of affiliation and (**b**) affiliation type.

3. Highlights

3.1. DRI-Based EAF Steelmaking

An important open question in DRI-based steelmaking is as follows: what is the most effective way to alloy carbon into the steel? Carbon can be introduced either through DRI carburization or by carburizing the melt in the EAF. Another challenge associated with the use of DRI is its potential reoxidation during transport and storage. It is therefore fitting that this Special Issue includes a review paper by Kieush et al. (Contribution 8), which provides a comprehensive discussion of the current understanding of these aspects. The study concludes that the presence of carbon in the DRI offers some benefits for processing in the EAF compared to carbon-alloying directly in the EAF. The briquetting of DRI was identified as one of the most effective countermeasures against reoxidation. Finally, it is noted that the literature on the reoxidation of H-DRI remains scarce, representing an important research gap that should be addressed in future studies.

H-DRI melting has recently been studied by many research groups. The aspects investigated aspects include melting in a high-temperature atmosphere [25,26], the modelling of melting in molten steel [27], experimental [28] and numerical [29] studies of melting in slag, the modelling of capillary infiltration of slag into pores [30], the experimental study of optical emissions during melting [31], and the characterization of the resulting slags [13]. In this Special Issue, the paper by Pfeiffer et al. (Contribution 7) investigates the differences in the melting behavior of carbon-free and carbon-containing DRI. Melting trials in a laboratory-scale EAF were conducted to compare carbon-free and carbon-containing DRI from DR-grade pellets as well as fines from a fluidized bed reactor. In a fourth trial, the influence of a slag layer was investigated by melting DRI from BF-grade pellets with the continuous addition of slag-forming oxides. The authors confirmed that feeding DRI into the EAF hot spot is important for the rapid melting of the material. Furthermore, the authors found that a higher carbon content in the DRI is beneficial for the melting stage, leading them to state that the optimal condition is *"a highly carburized DRI that is fed directly into the electric arc hot spot"*.

3.2. *Substitution of Fossil Carbon*

Fossil carbon is one of the main sources of CO_2 emissions in EAF steelmaking, with estimates indicating that its use leads to around 43 kg of CO_2 per ton of steel, corresponding to 40–70% of the total direct CO_2 emissions [32]. Consequently, the substitution of fossil carbon with non-fossil carbon materials has attracted interest, particularly in Europe. One potential application of biochar in an EAF is slag foaming [33]. In this Special Issue, Kieush et al. (Contribution 6) present a study on the thermochemical properties of biocoke to which wood pellets have been added, focusing on its potential applications for slag foaming in EAFs. It was found that biocoke supplemented with 5 wt% wood pellets could yield even better slag-foaming characteristics than conventional laboratory-scale coke, which was used as reference. At a higher wood pellet content of 10 wt%, the slag-foaming properties were slightly worse than those obtained with coke. However, due to their higher reactivity, the biocokes supplemented with wood pellets exhibited shorter foaming times. These results highlight the potential of biocoke for slag foaming and the need to carefully adjust its recipe.

The substitution of fossil carbon is also a topic in the valorization of steelmaking waste like mill scale. In this Special Issue, Dall'Osto et al. (Contribution 2) investigate how the physicochemical properties of biogenic carbon influence both the manufacturing and industrial transferability of mill-scale-based self-reducing briquettes. The authors systematically characterize several biogenic carbon candidates and mill scale and correlate their attributes with briquetting behavior (via impact resistance, cold compression, and water immersion resistance tests) and reduction performance. They show that the hydrophilicity or hydrophobicity of the reducing agents indirectly controls the mechanical performance of the briquettes produced. Furthermore, iron recovery is regulated by the amount of fixed carbon and volatile matter in the agglomerate. Building on these results, the authors formulate subsequent research topics to be addressed, including briquette charging conditions, the reduction behavior of briquettes in interaction with slag and metal, and the influence of briquette addition on steel quality and melting furnace productivity. Overall, this study reveals a viable pathway for recycling mill scale with biogenic reductants in melting furnaces like EAFs.

3.3. Modeling and Simulation

Substantial efforts have been dedicated to developing models that simulate different aspects of the EAF process [34–36]. In this Special Issue, modelling and simulation play a significant role in the studies by Azzaz et al. (Contribution 1), Reinicke et al. (Contribution 3), and Ugarte et al. (Contribution 5). The vastly different approaches described in these papers highlight the variety of methods available for numerical modelling and simulation.

Models based on machine learning approaches are particularly popular for endpoint prediction regarding off-gas composition, carbon content, metal bath temperature, energy consumption, etc. [35,37]. In this Special Issue, Azzaz et al. (Contribution 1) studied the use of machine learning models to predict the final P content in the scrap-based EAF process. Models based on random forest, support vector machine, and artificial neural network (ANN) approaches were tested. The ANN models employed the Adam optimizer and a nonlinear sigmoid activation function, with varying numbers of hidden layers. Among the models tested, the ANN model with four hidden layers proved to be the most effective. The results indicate that the final P content was highly correlated with the Cr and S content in scrap, injected oxygen, and process duration (p-value < 0.01) and moderately correlated with scrap weight, the Mn content of the scrap, and the amount of lime injected (0.01 < p-value < 0.05).

Process models are a loosely defined group of models. Generally, the aim is to describe the dynamics of a process based on known process variables. Phenomena-based process models are invariably based on mass and heat balances and are usually combined with simplified descriptions of local mass and heat transfer, without a spatial solution of momentum transfer as in computational fluid dynamics (CFD) models. Such models for the EAF process were recently reviewed by Hay et al. [36]. The dynamic process model developed at RWTH Aachen University, which is strongly influenced by the model developed by Logar et al. [38,39] at the University of Ljubljana, was first developed by Meier [40] and subsequently modified by Hay [41]. In this Special Issue, the paper by Reinicke et al. (Contribution 3) continues this line of modelling work by introducing an ANN-based surrogate model for the activity of slag species that was trained using simulation data from FactSage. More specifically, the activities were approximated with a shallow feedforward ANN consisting of an input layer (slag composition and temperature), an output layer (activities), and a single hidden layer, in which a rectified linear unit was used to achieve nonlinear transformation. The performance of the ANN-based surrogate model was benchmarked against that of the regular solution model (RSM) and the cell model (CM) implemented earlier by Hay [41]. While the comparison with FactSage simulations does not directly demonstrate the accuracy of the models relative to experimental data, it seems reasonable to assume that the FactSage simulations are sufficiently close to the experimental data for the EAF slag systems considered. As expected, the ANN-based surrogate model performed substantially better than RSM or CM both in terms of mean absolute error and coefficient of determination. The process model was somewhat slower when using the ANN-based surrogate model (0.115 s for 1000 samples) than when using RSM (0.037 s), but it was substantially faster than when CM was used (9.848 s). These results indicate that although the ANN-based surrogate model requires prior training with synthetic data, its substantial accuracy improvement over both RSM and CM, combined with its significant performance improvement over CM, makes it the preferred option in all cases except when the fastest possible execution speed is required or when the model is deliberately applied outside the validity range of the ANN-based surrogate model.

Ugarte et al. (Contribution 5) studied the impact of the oxygen injection rate on mixing using CFD simulations performed in ANSYS (Fluent version 2019). First, the supersonic coherent jet was simulated as a compressible, non-isothermal flow under steady-state conditions, providing velocity profiles and spatial variations in gas flow composition from the burner tip to the bath surface. In the second step, these outputs were used to estimate the cavities formed by the gas jets through empirical correlations. Finally, the authors constructed the geometry of the metal bath in an industrial EAF, incorporating the cavities calculated in the second step, and used it for transient refining simulations in the impact area domain. The authors found that assigning the highest flow rates to coherent jets positioned opposite each other across the furnace reduced the mixing time by about 10% compared to the baseline case.

3.4. System-Level Investigations

At the system level, Dall'Osto et al. (Contribution 4) analyze the consequences of creating a national direct reduction and electric steelmaking grid for Italy, moving beyond furnace-scale phenomena to the orchestration of assets, infrastructures, and markets. Starting from the current Italian production portfolio, which is strongly EAF-oriented for long products and still reliant on BF–BOF for flat products, they develop scenarios in which BF–BOF capacity is progressively replaced by DR–EAF. The scenarios address three different pathways, based on natural gas, green hydrogen, and biomethane as the main gas stream, and show how the emissions and energy demand of the Italian steel sector could change during the establishment of a DR–EAF grid. The authors found that although in principle each of the pathways could be feasible, the development of the production share of the DR–EAF grid in the future is vitally linked to the development of a national renewable energy policy and the generation of biomethane production capacity as well as meeting the ambitious 2050 targets for the Italian renewable energy grid.

4. Conclusions

Driven by the increasing availability of scrap and plans to replace BF-based ironmaking with DRI production, both scrap- and DRI-based steelmaking are projected to increase in the future. This trend not only highlights the importance of EAF research but also broadens its scope, particularly regarding the role of EAFs in direct reduction routes.

The eight peer-reviewed papers included in this Special Issue address many of the hot topics in current EAF research, such as DRI handling, DRI melting, the use of biocoke for slag foaming, and dephosphorization. The results presented highlight the suitability of biocoke supplemented with wood pellets for slag foaming, the usefulness of feeding DRI into the electric arc hot spot, factors governing the mechanical performance and iron recovery of mill-scale-based self-reducing briquettes, etc.

Regarding modelling and simulation, the boundaries between different approaches are beginning to blur, as machine learning methods are used both independently and to support other types of simulation activities. In process modelling, models can be accelerated using surrogate models to describe simulation results for properties with complex dependencies, such as the activities of species in the slag. A detailed CFD simulation of the entire EAF process—including heat transfer and chemical reactions—within an industrially reasonable timeframe remains a formidable challenge. However, the increasing availability of computational resources enables the use of detailed modelling approaches to simulate individual aspects of the process.

Author Contributions: Conceptualization, V.-V.V.; methodology, V.-V.V. and T.E.; writing—original draft preparation, V.-V.V.; writing—review and editing, V.-V.V. and T.E.; visualization, V.-V.V.; project administration, V.-V.V. and T.E. All authors have read and agreed to the published version of the manuscript.

Funding: V.-V.V. was supported by the CO_2-lean electric arc furnace steelmaking through fundamental and data-driven mathematical modelling (LEAF) project (no. 356439), funded by the Research Council of Finland, and by the Towards Carbon Neutral Metals 2 (TOCANEM 2), funded by Business Finland.

Data Availability Statement: No new data were created or analyzed in this study. Data sharing is not applicable to this article.

Acknowledgments: Rita Kallio, Ilpo Mäkelä, Saku Rytky, and Petri Sulasalmi from the University of Oulu are acknowledged for their fruitful discussions on various aspects of the EAF process.

Conflicts of Interest: The authors declare no conflicts of interest.

List of Contributions:

1. Azzaz, R.; Jahazi, M.; Ebrahimi Kahou, S.; Moosavi-Khoonsari, E. Prediction of Final Phosphorus Content of Steel in a Scrap-Based Electric Arc Furnace Using Artificial Neural Networks. *Metals* **2025**, *15*, 62.
2. Dall'Osto, G.; Mombelli, D.; Scolari, S.; Mapelli, C. Role of the Biogenic Carbon Physicochemical Properties in the Manufacturing and Industrial Transferability of Mill Scale-Based Self-Reducing Briquettes. *Metals* **2024**, *14*, 882.
3. Reinicke, A.; Engbrecht, T.-N.; Schüttensack, L.; Echterhof, T. Application of an Artificial Neural Network for Efficient Computation of Chemical Activities within an EAF Process Model. *Metals* **2024**, *14*, 736.
4. Dall'Osto, G.; Mombelli, D.; Mapelli, C. Consequences of the Direct Reduction and Electric Steelmaking Grid Creation on the Italian Steel Sector. *Metals* **2024**, *14*, 311.
5. Ugarte, O.; Busa, N.; Konar, B.; Okosun, T.; Zhou, C.Q. Impact of Injection Rate on Flow Mixing during the Refining Stage in an Electric Arc Furnace. *Metals* **2024**, *14*, 134.
6. Kieush, L.; Schenk, J.; Koveria, A.; Hrubiak, A. Biocoke Thermochemical Properties for Foamy Slag Formations in Electric Arc Furnace Steelmaking. *Metals* **2023**, *14*, 13.
7. Pfeiffer, A.; Ernst, D.; Zheng, H.; Wimmer, G.; Schenk, J. The Behavior of Direct Reduced Iron in the Electric Arc Furnace Hotspot. *Metals* **2023**, *13*, 978.
8. Kieush, L.; Lesiak, S.; Rieger, J.; Leitner, M.; Schmidt, L.; Daghagheleh, O. Reoxidation Behavior of the Direct Reduced Iron and Hot Briquetted Iron during Handling and Their Integration into Electric Arc Furnace Steelmaking: A Review. *Metals* **2024**, *14*, 873.

References

1. Karbowniczek, M. *Electric Arc Furnace Steelmaking*; CRC Press: Boca Raton, FL, USA, 2021.
2. Cappel, J.; Ahrenhold, F.; Egger, M.W.; Hiebler, H.; Schenk, J. 70 Years of LD-Steelmaking—Quo Vadis? *Metals* **2022**, *12*, 912. [CrossRef]
3. Madias, J. Electric Furnace Steelmaking. In *Treatise on Process Metallurgy*, 2nd ed.; Volume 3: Industrial Processes; Seetharaman, S., Guthrie, R., McLean, A., Seetharaman, S., Sohn, H.Y., Eds.; Elsevier: Amsterdam, The Netherlands, 2024; pp. 243–265.
4. Klein, K.-H.; Schindler, J.E. Metallurgie bei Schrotteinsatz. In *Elektrostahl-Erzeugung*; Heinen, K.-H., Ed.; Verlag Stahleisen GmbH: Düsseldorf, Germany, 1997; pp. 473–502.
5. Walden, K. Metallurgie bei Eisenschwammeinsatz. In *Elektrostahl-Erzeugung*; Heinen, K.-H., Ed.; Verlag Stahleisen GmbH: Düsseldorf, Germany, 1997; pp. 503–511.
6. World Steel Association. *World Steel in Figures 2025*; World Steel Association: Beijing, China, 2025.
7. Ariyama, T. Perspectives on the Promising Pathways to Zero Carbon Emissions in the Steel Industry toward 2050. *ISIJ Int.* **2025**, *65*, 165–184. [CrossRef]
8. Pei, M.; Petäjäniemi, M.; Regnell, A.; Wijk, O. Toward a Fossil Free Future with HYBRIT: Development of Iron and Steelmaking Technology in Sweden and Finland. *Metals* **2020**, *10*, 972. [CrossRef]
9. Blastr Green Steel. Blastr Green Steel Chooses Primetals Technologies as Its Technological Partner for the Ultra-Low CO_2 Emissions Steel Plant in Inkoo, Finland. *Blastr Green Steel*, 9 July 2024.
10. SMS Group. The World's First Carbon-Neutral Steel Plant. Available online: https://www.sms-group.com/plants/projects/stegra (accessed on 11 November 2025).
11. SSAB. The Deputy Prime Minister and SSAB's CEO Broke Ground on a New Steel Mill in Luleå. *SSAB*, 17 September 2025.

12. Hassan, A.; Kotelnikov, G.; Abdelwahed, H. Melting Characteristics of Alternative Charging Materials in an Electric Arc Furnace Steelmaking. *Ironmak. Steelmak.* **2021**, *48*, 1136–1141. [CrossRef]
13. Kallio, R.; Cantaluppi, M.; Louhisalmi, J.; Visuri, V.-V. Mineralogical Characteristics of Fossil-Free Steel Slags. *Miner. Eng.* **2025**, *230*, 109396. [CrossRef]
14. Wimmer, G.; Voraberger, B.; Kradel, B.; Fleischanderl, A. Breakthrough Pathways to Decarbonize the Steel Sector. *Mitsubishi Heavy Ind. Tech. Rev.* **2022**, *59*, 1–7.
15. Wimmer, G.; Voraberger, B.; Rosner, J.; Pfeiffer, A. Smelter—Green Steelmaking Using Low-Grade DRI. In Proceedings of the Iron & Steel Technology Conference, Columbus, OH, USA, 6–9 May 2024; pp. 180–187.
16. Metso. Metso Opens DRI Smelting Furnace Pilot Facility in Pori, Finland. *Metso*, 25 October 2024.
17. Weinberg, M. Transformation of Steel Production. In Proceedings of the Plenary presentation—13th European Electric Steelmaking Conference, Essen, Germany, 3–7 June 2024; Steel Institute VDEh: Essen, Germany, 2024.
18. Worldstainless. Stainless Steel Melt Shop Production Increases by 7% in 2024. *Worldstainless*, 14 April 2025.
19. Visuri, V.-V.; Holappa, L. Converter Steelmaking. In *Treatise on Process Metallurgy—Volume 3: Industrial Processes*; Seetharaman, S., Guthrie, R., McLean, A., Seetharaman, S., Sohn, H.Y., Eds.; Elsevier: Amsterdam, The Netherlands, 2024; pp. 183–241.
20. Satritama, B.; Cooper, C.; Fellicia, D.; Pownceby, M.I.; Palanisamy, S.; Ang, A.; Mukhlis, R.Z.; Pye, J.; Rahbari, A.; Brooks, G.A.; et al. Hydrogen Plasma for Low-Carbon Extractive Metallurgy: Oxides Reduction, Metals Refining, and Wastes Processing. *J. Sustain. Metall.* **2024**, *10*, 1845–1894. [CrossRef]
21. Adami, B.; Hoffelner, F.; Zarl, M.A.; Schenk, J. Strategic Selection of a Pre-Reduction Reactor for Increased Hydrogen Utilization in Hydrogen Plasma Smelting Reduction. *Processes* **2025**, *13*, 420. [CrossRef]
22. Zarl, M.A.; Farkas, M.A.; Schenk, J. A Study on the Stability Fields of Arc Plasma in the HPSR Process. *Metals* **2020**, *10*, 1394. [CrossRef]
23. Büyükuslu, Ö.K.; Aota, L.S.; Raabe, D.; Springer, H.; Souza Filho, I.R. Mechanisms and Elemental Partitioning during Simultaneous Dephosphorization and Reduction of Fe-O-P Melts by Hydrogen Plasma. *Acta Mater.* **2024**, *277*, 120221. [CrossRef]
24. Sassi, E.-M. Slag and Refractory Interactions in Hydrogen Plasma Smelting Reduction. Master's Thesis, University of Oulu, Oulu, Finland, 2025.
25. Huss, J.; Vickerfält, A.; Kojola, N. Some Aspects of the Melting and Dephosphorization Mechanism of Hydrogen-DRI. *Steel Res. Int.* **2023**, *94*, 2300064. [CrossRef]
26. Vickerfält, A.; Huss, J.; Martinsson, J.; Sichen, D. Reaction Mechanisms During Melting of H-DRI Focusing on Slag Formation and the Behavior of Vanadium. *Metall. Mater. Trans. B* **2023**, *54*, 2206–2215. [CrossRef]
27. Govro, J.; Meena, A.; Chakraborty, S.; Meshram, A.; Korobeinikov, Y.; Phillips, K.; Athavale, V.; Bartlett, L.; Smith, J.; Emdadi, A.; et al. Melting Behavior of Hydrogen-Reduced DRI in a Simulated EAF Steel Bath. *Iron Steel Technol.* **2023**, *20*, 58–63.
28. Huss, J.; Vickerfält, A.; Kojola, N. The Melting Mechanism of Hydrogen Direct Reduced Iron in Liquid Slag. *Steel Res. Int.* **2024**, *95*, 2300325. [CrossRef]
29. Svantesson, J.; Kojola, N.; Ersson, M. Numerical Study on the Effect of Material Parameters and Process Conditions on the Melting Time of Hydrogen-Direct Reduced Iron. *Metall. Mater. Trans. B* **2025**, *56*, 2846–2872. [CrossRef]
30. Svantesson, J.L.; Ersson, M.; Kojola, N. Capillary Infiltration of Slag in Hydrogen-Direct Reduced Iron and Influence on Melting. *ISIJ Int.* **2025**, *65*, 1607–1619. [CrossRef]
31. Pauna, H.; Kokkonen, T.; Cavaliere, P.; Bayat, M.; Mirowska, A.; Alatarvas, T.; Huttula, M.; Fabritius, T. Hydrogen Direct Reduced Iron Melting in an Electric Arc Furnace: Benefits of In Situ Monitoring. *J. Sustain. Metall.* **2025**, *11*, 1–13. [CrossRef]
32. Demus, T.; Echterhof, T.; Pfeifer, H.; Schulten, M.; Noel, Y.; Quicker, P. Investigations on the Use of Biogenic Residues as a Substitute for Fossil Coal in the EAF Steelmaking Process. In Proceedings of the 10th European Electric Steelmaking Conference, Graz, Austria, 25–28 September 2012; ASMET: Graz, Austria, 2012.
33. Liu, X.; Yan, W. Current Advances in Slag Foaming Processes toward Reduced CO_2 Emission for Electric Arc Furnace Steelmaking. *J. CO2 Util.* **2024**, *90*, 102979. [CrossRef]
34. Odenthal, H.-J.; Kemminger, A.; Krause, F.; Sankowski, L.; Uebber, N.; Vogl, N. Review on Modeling and Simulation of the Electric Arc Furnace (EAF). *Steel Res. Int.* **2018**, *89*, 1700098. [CrossRef]
35. Carlsson, L.S.; Samuelsson, P.B.; Jönsson, P.G. Predicting the Electrical Energy Consumption of Electric Arc Furnaces Using Statistical Modeling. *Metals* **2019**, *9*, 959. [CrossRef]
36. Hay, T.; Visuri, V.-V.; Aula, M.; Echterhof, T. A Review of Mathematical Process Models for the Electric Arc Furnace Process. *Steel Res. Int.* **2021**, *92*, 2000395. [CrossRef]
37. Niyayesh, M.; Yilmaz, U. Predicting Endpoint Parameters of Electric Arc Furnace–Based Steelmaking Using Artificial Neural Network. *Int. J. Adv. Manuf. Tech.* **2025**, *138*, 155–167. [CrossRef]
38. Logar, V.; Dovžan, D.; Škrjanc, I. Modeling and Validation of an Electric Arc Furnace: Part 1, Heat and Mass Transfer. *ISIJ Int.* **2012**, *52*, 402–412. [CrossRef]

39. Logar, V.; Dovžan, D.; Škrjanc, I. Modeling and Validation of an Electric Arc Furnace: Part 2, Thermo-Chemistry. *ISIJ Int.* **2012**, *52*, 413–423. [CrossRef]
40. Meier, T. Modellierung und Simulation des Elektrolichtbogenofens. Doctoral Thesis, RWTH Aachen University, Aachen, Germany, 2016.
41. Hay, T.A. Mathematische Modellierung des Elektrostahlverfahrens. Doctoral Thesis, RWTH Aachen University, Aachen, Germany, 2021.

metals

MDPI

Article

Prediction of Final Phosphorus Content of Steel in a Scrap-Based Electric Arc Furnace Using Artificial Neural Networks

Riadh Azzaz [1], Mohammad Jahazi [1], Samira Ebrahimi Kahou [2] and Elmira Moosavi-Khoonsari [1,*]

[1] Department of Mechanical Engineering, École de Technologie Supérieure (ÉTS), 1100 Notre-Dame Street West, Montréal, QC H3C 1K3, Canada; riadh.azzaz.1@ens.etsmtl.ca (R.A.); mohammad.jahazi@etsmtl.ca (M.J.)
[2] Schulich School of Engineering, Department of Electrical and Software Engineering, University of Calgary, 856 Campus Pl NW, Calgary, AB T2N 4V8, Canada; samira.ebrahimikahou@ucalgary.ca
* Correspondence: elmira.moosavi@etsmtl.ca

Abstract: The scrap-based electric arc furnace process is expected to capture a significant share of the steel market in the future due to its potential for reducing environmental impacts through steel recycling. However, managing impurities, particularly phosphorus, remains a challenge. This study aims to develop a machine learning model to estimate steel phosphorus content at the end of the process based on input parameters. Data were collected over one year from a steel plant, focusing on parameters such as the chemical composition and weight of the scrap, the volume of oxygen injected, injected lime, and process duration. After preprocessing the data, several machine learning models were evaluated, with the artificial neural network (ANN) emerging as the most effective. The Adam optimizer and non-linear sigmoid activation function were employed. The best ANN model included four hidden layers and 448 neurons. The model was trained for 500 epochs with a batch size of 50. The model achieves a mean square error (MSE) of 0.000016, a root mean square error ($RMSE$) of 0.0049998, a coefficient of determination (R^2) of 99.96%, and a correlation coefficient (r) of 99.98%. Notably, the model was tested on over 200 unseen data points and achieved a 100% hit rate for predicting phosphorus content within $\pm$0.001 wt% ($\pm$10 ppm). These results demonstrate that the optimized ANN model offers accurate predictions for the steel final phosphorus content.

Keywords: steelmaking; scrap-based electric arc furnace; artificial neural network; machine learning; dephosphorization

1. Introduction

Steelmaking is currently a major contributor to CO_2 emissions, but it is committed to advancing a sustainable metallurgical industry, as reflected in its adoption of scrap-based electric arc furnaces (EAFs). This process involves melting scrap steel by generating an electric arc between electrodes and the liquid steel bath [1]. It effectively recycles steel scrap, reducing CO_2 emissions by 90% and energy consumption by 70% compared to the traditional blast furnace–basic oxygen furnace (BF-BOF) route. Additionally, it significantly lowers the consumption of natural resources like iron ore, coal, and limestone [2,3].

Despite the environmental benefits of EAFs, they face complex scientific and technical challenges, particularly in managing impurities such as phosphorus (P) in steel [4,5]. An uncontrolled quantity of P in steel negatively impacts the mechanical properties of steel, leading to increased temper and intergranular embrittlement and cracking [6,7]. To meet quality standards, it is crucial to reduce P levels from typically above 0.025 wt% in scrap to less than 0.015 wt% in the final product. For certain applications, the target phosphorus content may need to be as low as 0.005 wt% [8]. The varied composition of scrap feedstock in comparison to ore-based production further complicates this reduction process, and steelmaking needs to align its operation continuously with the complicated composition of modern steel products [9–11].

Numerous studies have investigated P removal from steel, focusing on P equilibrium distribution and phosphate capacity at laboratory or intermediate scales [12–26]. Additionally, plant trials in EAFs have examined P behavior during direct reduced iron (DRI) and hot briquetted iron (HBI) processes [27–29]. While experimental methods are valuable, they are often time-consuming, costly, and difficult to apply on an industrial scale. Furthermore, P measurements in controlled lab conditions do not easily translate to large-scale environments where fluid flow and kinetic conditions differ significantly. Consequently, modeling and simulation provide viable alternatives to purely experimental methods, primarily divided into phenomenological (mechanistic) models based on physical phenomena, such as computational fluid dynamics (CFD) [30,31], and data-driven statistical models [32].

Mechanistic models have greatly enhanced our understanding of the EAF process, but they come with limitations. For example, CFD models include reliance on equilibrium models for metal–slag–gas interactions, which sacrifice accuracy for speed, and statistical turbulence modeling that may introduce errors in unsteady flow conditions. Additionally, the use of empirical constants for mass transfer coefficients limits generalizability, while inadequate validation of foamy slag models and the oversimplification of local conditions reduce overall accuracy. The assumption of arc plasma as a black body for heat transfer may also be an oversimplification, and the computational demands of comprehensive models make them impractical for online applications [30,32]. Consequently, while phenomenological models show promise, they still face challenges, particularly in capturing the wide range of scales and phenomena in such a complex EAF process.

In light of the limitations of traditional physical models, researchers have increasingly turned to statistical and data-driven approaches for predicting the final P content in steel [33–48]. The machine learning (ML) methods offer a faster, cheaper, and safer alternative to plant trials [32] and adapt well to variations in scrap composition and operating conditions, often outperforming mechanistic models in terms of accuracy [32,46].

While only a few studies have focused on predicting and optimizing the EAF process [34,45,47,49–51], particularly regarding endpoint P content [34,44,47], most research has concentrated on the BOF process [33,35–39,41–43,45,46,52]. As EAFs gain prominence and scrap recycling becomes crucial, accurate P prediction is vital. Existing ML models show promise, but their effectiveness varies due to data availability and quality, input parameter selection, and model robustness [45,53]. Notably, Yuan et al. [34] developed a least squares support vector machine (LS-SVM) model that achieved an 87% hit rate for predicting P levels with ±0.003 wt% errors in EAF steel. Chen et al. [44] developed a back propagation neural network–decision tree (BPNN-DT) model with six hidden layers, 18 input parameters, and 50 neurons to predict the final P content of steel in EAF. The proposed hybrid model combines k-means clustering, BPNN, and a DT algorithm for prediction. The model achieves a phosphorous prediction accuracy of 83.0% for ±0.004 wt% error range. Zou et al. [47] used a BPNN model with 14 hidden layers, attaining a hit rate of 87.8% for ±0.004 wt% errors and 75.6% for ±0.003 wt% errors in phosphorous prediction. In industrial contexts, 20% of wrong predictions might still result in inefficiency, waste, or poor-quality products, which could affect the overall performance of the system. Increasing the hit rate improves the model's accuracy, leading to more precise decision-making, reduced mistakes, and enhanced operational efficiency, and builds confidence in using the model for process optimization.

This study aims to develop an ML model to predict the ultimate phosphorus content of steel with higher accuracy, using key input parameters from a scrap-based EAF process. While previous studies have focused on similar predictions, this work is the first to develop a model specifically for EAF processes operating exclusively with scrap, and it is also the first to consider the composition of the scrap. The approach includes preprocessing original production data from a steelmaking plant to remove outliers and performing a correlation analysis between input parameters and the phosphorous content. An ANN model is compared with random forest (RF), SVM with a radial basis function (RBF) kernel, and models from the literature using various evaluation metrics to assess the predictive performance.

2. Analysis of Scrap-Based EAF

2.1. Description of EAF Process

An EAF operates in batch tap-to-tap cycles, consisting of the following steps: initial charging (3 min), primary melting (20 min), additional charging (3 min), secondary melting (14 min), refining (10 min), deslagging and tapping (3 min), and furnace tilting (7 min). Modern operations aim to complete the entire tap-to-tap cycle in under 60 min [54]. A schematic of EAF steelmaking is shown in Figure 1.

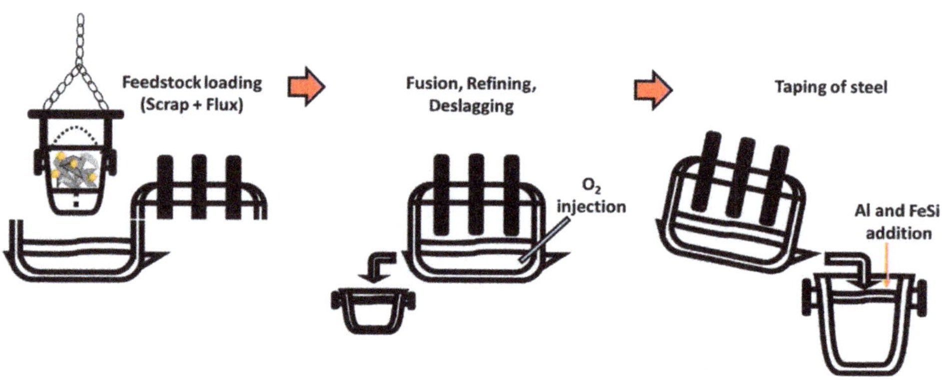

Figure 1. Schematic of an electric arc furnace steelmaking process.

Charging the Furnace. The furnace is charged from the top and many companies combine lime and carbon addition in the scrap basket and use additional injections as needed [55]. The number of scrap buckets used is based on furnace volume and scrap density, with modern designs aiming to minimize recharging to reduce downtime and energy loss. Typically, companies aim for two to three scrap buckets per cycle [54].

Melting Scrap. Melting scrap in an EAF primarily relies on electrical energy supplied by graphite electrodes. Initially, an intermediate voltage is used until the electrodes penetrate the scrap, after which a higher voltage stabilizes the arc for efficient heat transfer and forms a liquid metal pool. Chemical energy, provided by oxy-fuel burners and oxygen lances, further aids the melting process through flame radiation, convection, and exothermic reactions. The process continues with repeated charging until all scrap is melted [56].

Refining. Once the bath temperature stabilizes, chemical analysis directs refining operations such as oxygen blowing and alloy additions. Oxygen injection begins before stabilization, initiating some reactions early. Adjustments are made to manage excess elements like phosphorous, carbon, silicon, and chromium by transferring them to the slag phase. However, the EAF's impurity removal capacity is limited due to the lower basicity and mass of the slag. Initial slagging is crucial for removing phosphorous before reversion occurs. The final bath composition is carefully managed to meet steel specifications, with alloy additions made in the ladle to adjust the composition as needed [42,54,56,57].

Deslagging. The slag collecting the undesired species like phosphorous is removed during the deslagging step by tilting the furnace backward and allowing the slag to exit through a designated door. This removal process reduces the risk of phosphorous reversion when the temperature is increased for further refining, such as during desulfurization or carbon injection, and during slag foaming to reduce iron oxide to metallic iron [54,58].

Tapping. Tapping molten metal from a furnace is a crucial operation, and any failure in this process necessitates a complete shutdown. Operations can only resume once tapping is successfully completed. Key factors to manage during tapping are the rate and duration. It is also important to note that the furnace is never entirely emptied; a small amount of molten metal remains inside when the tapping hole is sealed [59].

2.2. Phosphorous Removal

The phosphorous removal process can be divided into two main stages [60]. Initially, P in iron-based melts is oxidized by Fe_tO, which is primarily generated from the reaction of scrap with injected oxygen, forming P_2O_5 according to the following reaction, as shown in Equation (1):

$$2[\mathrm{P}] + 5(\mathrm{Fe_tO}) = (\mathrm{P_2O_5}) + 5\mathrm{t}[\mathrm{Fe}] \tag{1}$$

where [ ] and ( ) denote the species in the metal and slag phases, respectively.

Next, the injected flux (CaO) stabilizes the extracted phosphorus (P_2O_5) in the slag, resulting in the formation of calcium phosphate ($3CaO{\cdot}P_2O_5$) through the following reaction, as shown in Equation (2):

$$(\mathrm{P_2O_5}) + 3(\mathrm{CaO}) = (3\mathrm{CaO}\ \mathrm{P_2O_5}) \tag{2}$$

The phosphorus removal reaction can also be represented in its ionic form, as shown in Equation (3) [25]:

$$[\mathrm{P}] + \frac{5}{2}[\mathrm{O}] + \frac{3}{2}(\mathrm{O^{2-}}) = (\mathrm{PO_4^{3-}}) \tag{3}$$

where [P] and [O] represent phosphorus and oxygen, respectively, and O^{2-} and PO_4^{3-} represent the oxide and phosphate ions, respectively.

Two concepts, phosphorus partition coefficient (L_p) and phosphate capacity ($C_{PO_4^{3-}}$), have been developed to quantify the phosphorus removal process [25]. The L_p parameter can be described as follows:

$$L_p = \frac{(\%\mathrm{P})}{[\%\mathrm{P}]} \tag{4}$$

where (%P) and [%P] represent the phosphorus concentrations in the slag and steel, respectively. The L_p parameter ranges from 5.0 to 15.0. Generally, phosphorus content is only reduced by about 20 to 50% during EAF treatment. However, given the low phosphorus content of scrap compared to hot metal (produced from iron ore treatment in the BF), this degree of removal is considered satisfactory [54].

The L_p parameter between the slag and liquid steel is commonly used to evaluate the phosphorus removal capability of the slag due to its ease of measurement in both laboratory studies and commercial production [12–26,60,61]. Nevertheless, it is crucial to note that this ratio can only be used as a comparative measure between different slag compositions if the partial oxygen pressure (P_{O_2}) is equivalent in the compared systems [62].

Wagner proposed the concept of phosphate capacity ($C_{PO_4^{3-}}$) to describe the slag's phosphorus removal potential using a slag–gas equilibrium reaction. $C_{PO_4^{3-}}$ incorporates the influence of P_{O_2}, making it an essential measure for the comparative evaluation of various slag systems. The slag–gas reaction and $C_{PO_4^{3-}}$ are represented by Equations (5) and (6), respectively [62]:

$$\frac{1}{2}P_2(g)+\frac{5}{4}O_2(g)+\frac{3}{2}O^{2-}\ (slag) = PO_4^{3-}\ (slag) \tag{5}$$

$$C_{PO_4^{3-}} = \frac{\left(\%PO_4^{3-}\right)}{P_{P_2}^{1/2}P_{O_2}^{5/4}} = \frac{K_{(2)}\left(a_{O^{2-}}\right)^{3/2}}{\gamma^{\circ}_{PO_4^{-3}}} \tag{6}$$

where $\%PO_4^{3-}$ is the weight percentage of PO_4^{3-} dissolved in the slag, and P_{O_2} and P_{P_2} are the partial pressures of oxygen and phosphorus, respectively, at the slag–gas interface in equilibrium. In cases where the concentration of PO_4^{3-} is notably low, it is acceptable to replace the activity with the corresponding concentrations of $(\%PO_4^{3-})$ multiplied by a critical constant parameter $\gamma^{\circ}_{PO_4^{-3}}$, representing the activity coefficient at infinite dilution. $K_{(2)}$ is the equilibrium constant for reaction (5).

$C_{PO_4^{3-}}$ shows a direct correlation with L_p as shown in Equation (7):

$$C_{PO_4^{3-}} = \frac{L_p k_p}{f_p P_{O_2}^{5/4}} \tag{7}$$

where k_p is the equilibrium constant for phosphorus dissolution in iron ($\frac{1}{2}P_{2(g)} = [P]$) [62].

2.3. Factors Influencing Phosphorus Removal

Several factors influence the effectiveness of phosphorus removal in the EAF. Key parameters affecting phosphorus elimination include the slag's basicity, temperature, and FeO content. From a thermodynamic point of view, low temperatures, high FeO content, and increased basicity generally favor the phosphorus removal process [19–22,63,64].

Basicity. The basicity of slag is usually expressed as the weight ratio of basic oxides (e.g., CaO) to acidic oxides (e.g., SiO_2). It is a critical factor in metallurgy, influencing the slag's ability to absorb impurities like phosphorus, as well as its melting point and viscosity [65,66]. Increasing the basicity of the slag, typically by raising the concentration of basic oxides such as CaO, enhances phosphorus removal efficiency by stabilizing P_2O_5 as $4CaO{\cdot}P_2O_5$ at steelmaking temperatures. However, overly high basicity can be counterproductive. Excessive basicity raises the slag's melting point, preventing complete melting of CaO particles and increasing slag viscosity. This increased viscosity reduces the phosphorus diffusion in the slag, slowing the phosphorus removal reaction at the interface between the molten steel and slag and thus diminishing removal efficiency [25,38,63].

FeO Content. The effectiveness of phosphorus removal in CaO-based slags is also influenced by the presence of iron(II) oxide (FeO), which can act as an acidic or basic oxide depending on the slag composition and oxygen potential. Research has shown that the phosphorus removal capacity, or phosphate capacity, of CaO-SiO_2-MgO-FeO slags increases with FeO content. Lee and Fruehan [20] observed an increase in the phosphate capacity with FeO content between 3 and 10 wt% at high temperatures. Hamano and Tsukihashi [19] found a maximum phosphate capacity at about 50 wt% FeO which then decreases when further increasing the FeO content to 60 wt%. Li et al. [21] noted that the phosphate capacity peaks at 25–35 wt% FeO and then decreases, which is attributed to the dilution of CaO, reducing its activity and increasing the activity of P_2O_5 [63]. Thus, optimizing FeO content is crucial for effective phosphorus removal in CaO-based slags.

Temperature. Temperature impacts phosphorus removal in two contrasting ways. High temperatures can negatively affect the process because phosphorus removal is highly exothermic. Conversely, elevated temperatures promote the melting of lime, which enhances the basicity of the slag. This improved basicity aids in the distribution of phosphorus into the slag phase and increases the L_p, thereby enhancing removal efficiency. On the other hand, temperature favors the kinetics of the phosphorus removal process [38,48].

3. Prediction of Endpoint Phosphorus Content in Steel

3.1. Machine Learning Algorithms

A wide range of ML models have been employed in steel dephosphorization for process prediction and optimization, such as different NNs, SVM, RF, gradient boosting regression (GBR), least squares SVM with principal component regression (LS-SVM-PCR), k-means-NN with decision tree (k-means-BPNN-DT), ridge regression, convolutional neural network (CNN), extreme learning machine (ELM), partial least squares (PLS), support vector regression (SVR), graph convolutional network (GCN), and general regression NN (GRNN), with various datasets ranging from small to large, and incorporating different numbers of input parameters to improve accuracy and efficiency in modeling steel production processes [34,36,37,41,43,44,46–48,51,52,63,67].

In this study, three specific regression models are employed: RF, SVM-RBF, and ANN. In general, SVM and RF are simpler than ANN and require less training time, which is why they were tested first. However, due to the need for higher prediction accuracy, we ultimately developed an ANN model, as the accuracy of the former models was insufficient. A detailed presentation of these techniques will be provided in the following section. The libraries used for developing the ANN, RF, and SVM models, along with their respective versions, are provided in the Appendix A.

3.1.1. Random Forest

Random forest was employed in this work for its prediction accuracy and its robustness in preventing overfitting. The model was implemented using the scikit-learn library version 1.5.2 with key hyperparameters, such as the number of trees and the criteria for splitting nodes (e.g., minimum node size). Tuning these hyperparameters can optimize performance; however, the RF model generally performs well with default settings provided in software packages [68].

3.1.2. Support Vector Machine

Support vector machines are renowned for their strong generalization capabilities and high prediction accuracy [69]. Key hyperparameters include the kernel type, gamma (γ), and regularization parameter (C), optimized to balance model complexity and prediction accuracy. The radial basis function (RBF) kernel was utilized in this work, the default choice, which is effective for modeling non-linear relationships [70,71].

3.1.3. Artificial Neural Network

Artificial neural networks are a robust ML framework known for their ability to model complex non-linear relationships, which is the case in metallurgical processes. A basic neural network consists of three main components: an input layer that receives data, an output layer that makes predictions, and one or more hidden layers that process information through interconnected neurons. Each neuron in the hidden layers operates with weights and biases, as described by Equation (8):

$$a = f\left[\sum_{i=1}^{n} w_i + b\right] \tag{8}$$

where w_i and b represent the weights and bias values, respectively, while x_i denotes the inputs and $f[.]$ denotes the activation function. Figure 2 illustrates an example ANN architecture and a basic neuron.

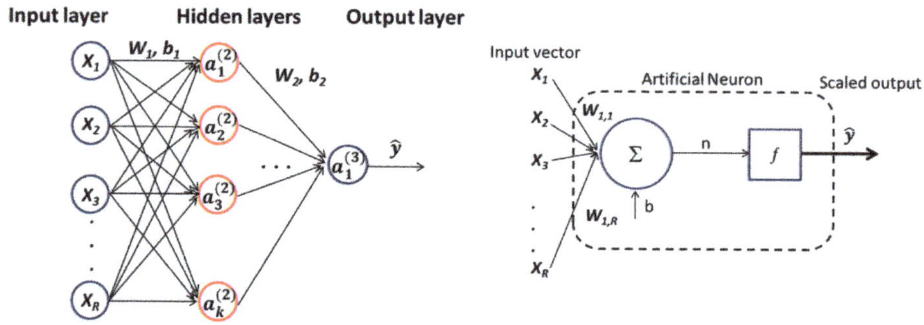

Figure 2. Example of an artificial neural network architecture along with a basic neuron.

During training, the network learns and adjusts the weights to optimize performance. The use of activation functions enables ANNs to learn complex patterns, making them capable of universal approximation—mapping any input to any output regardless of data complexity [72].

Establishment of Artificial Neural Network Models

Figure 3 illustrates the steps in developing an ANN model for this study. The process begins with data collection and preprocessing, including tasks such as data cleaning, correlation analysis, and normalization. The data are then split into training, validation, and test sets. Finally, the model is trained by selecting an appropriate architecture and fine-tuning key hyperparameters, such as the number of layers, neurons per layer, and the activation function.

Three categories of datasets are used: training, validation, and test sets. The training set provides information on the target function to train the network. The validation set is used in conjunction with early stopping techniques to monitor and prevent overfitting by tracking validation errors during training. After training, the test set is employed to evaluate the model's performance. Typically, 60% of the data are allocated for training, 20% for validation, and 20% for testing the model. The validation set is used strictly for monitoring model performance and tuning hyperparameters during training, while the test set is reserved for the final evaluation to ensure unbiased results.

The number of hidden layers and nodes within these layers is crucial for determining the performance of an ANN model. In this study, three different ANN architectures were tested, each varying in the number of hidden layers and nodes. All architectures employed the sigmoid activation function, a non-linear function by default, as shown in Equation (9). These models were implemented using the TensorFlow library version 2.18.0. The mean squared error (*MSE*) was used as the loss function for training, as detailed in Equation (10):

$$f(y) = \frac{1}{1+e^{-y}} \tag{9}$$

$$Loss\ (y,\ \hat{y}) = \frac{1}{N}\sum_{j=1}^{N}\left(y_j - \hat{y}_j\right)^2 \tag{10}$$

where $\hat{y}$ represents the predicted value, and y denotes the actual value.

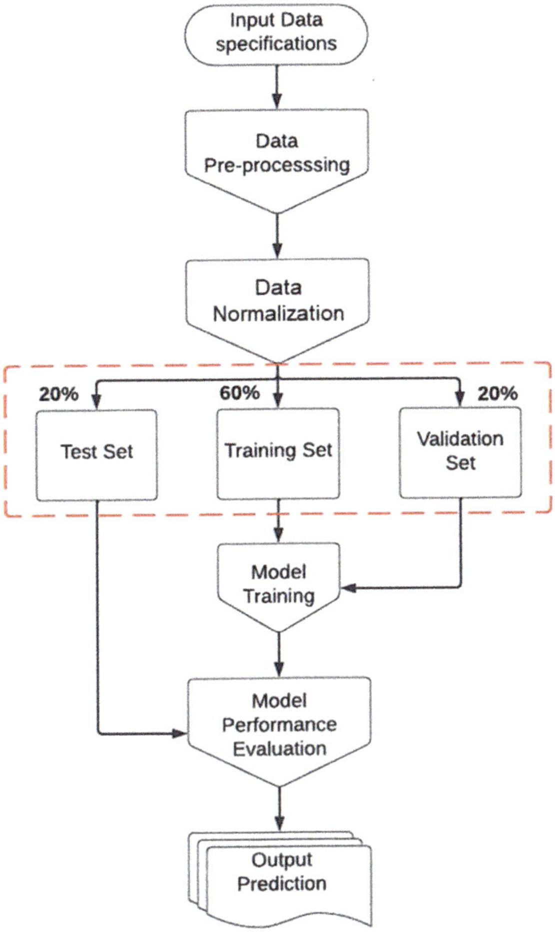

Figure 3. Flow chart for the development of the artificial neural network model based on historical plant data.

To optimize hyperparameters in an ANN model, three approaches have been proposed: grid search, random search, and manual trial and error [73,74]. In this work, the choice of ANN hyperparameters has been moderately searched to adapt to a low-data regime. In combination with our previous work [50], we tested configurations ranging from two to seven hidden layers and 24 to 464 neurons to identify the settings that yielded the lowest validation loss, measured by *MSE*. While each optimization technique has its advantages and limitations, trial and error can serve as a viable alternative to more advanced adaptive (sequential) hyperparameter optimization algorithms [74]. This method offers quick feedback and is straightforward to implement, requiring no prior expertise in complex optimization methods.

During training, the Adam optimizer was used to adaptively adjust learning rates for each parameter, enabling faster convergence and improved robustness to variations in the training data. The choice of Adam optimizer is default as it is accepted by the ML community to be the best optimizer for simple ML problems, as in our case, compared to traditional methods like stochastic gradient descent (SGD), which uses a fixed learning rate.

3.2. *Data Treatment*

The present study focuses on a 40-ton EAF equipped with three graphite electrodes and charged with two scrap bins. Initially, scrap from the first bin is melted in the superheated furnace, followed by the addition of scrap from the second bin. Chemical analysis of the steel is conducted at two critical stages: before deslagging and just before transferring the liquid metal to the ladle furnace (LF) at 1650 °C. The second analysis is crucial because, with the slag removed, phosphorus may revert into the steel, making final phosphorus content a key parameter for process control.

The steelmaking process produces a large volume of data, but these raw data often contain missing values, outliers, and inconsistencies, which can significantly impact model performance if used directly. Thus, preprocessing is essential to refine and prepare the data for ML. The methodology varies based on the quality and nature of the raw data. The following section will provide an overview of the data preprocessing stages used in the present study.

3.2.1. Data Collection

In this study, over 1700 heat datasets were collected from a steel plant over one year. These datasets include a range of variables, such as the chemical composition of scrap and various process parameters. In this EAF, the mass and composition of each scrap type are monitored, and often, no additional carbon is added. The overall composition of each steel was constructed based on the collected data in this work. Table 1 outlines the parameters used to develop the ANN models, including their symbols and the rationale for their selection. Twelve parameters were chosen based on principles of metallurgy, thermodynamics, and current industrial practices [19–22,63,64,75–79]. These parameters include the weight and composition of the scrap (C, Mn, Cr, Si, and S), the quantities of injected oxygen and lime, energy consumption, deslagging and tapping temperatures, and process duration.

Table 1. Input parameters used to develop the machine learning models.

Variables	Description of Variables	Justification
x_1	Scrap weight	Main material of EAF (source of P)
x_2	C content in scrap	
x_3	Mn content in scrap	
x_4	Cr content in scrap	Elements in scrap affecting dephosphorization
x_5	Si content in scrap	
x_6	S content in scrap	
x_7	Injected oxygen	Oxidant
x_8	Injected lime	Dephosphorization agent
x_9	Energy consumption	
x_{10}	Deslagging temperature	Process parameters
x_{11}	Tapping temperature	
x_{12}	Process duration	

3.2.2. Data Cleaning

In this work, data cleaning was performed after data collection to address issues such as missing or aberrant values. Understanding the distribution of the data, including central tendency, dispersion, and potential outliers, is crucial for making informed decisions. Box plots are used to graphically represent these characteristics in the case of processes where strong variability is observed, and the raw data were analyzed using the box plot concept, as shown in Figure 4. This method, as described by Dovoedo and Chakraborti [80], is employed to identify outliers. This method relies on four key statistics: the first quartile

(Q1), the median (Q2), the third quartile (Q3), and the interquartile range (IQR), which is the difference between Q3 and Q1. Outliers are defined as values falling below Q1 − 1.5 IQR or above Q3 + 1.5 IQR. In this study, outliers were removed based on these criteria. Figure 5 illustrates the data distribution in this work. For clarity in visualization, the data are presented on a scale from 0 to 1 in the box plot diagram, as the parameters have different scales and units.

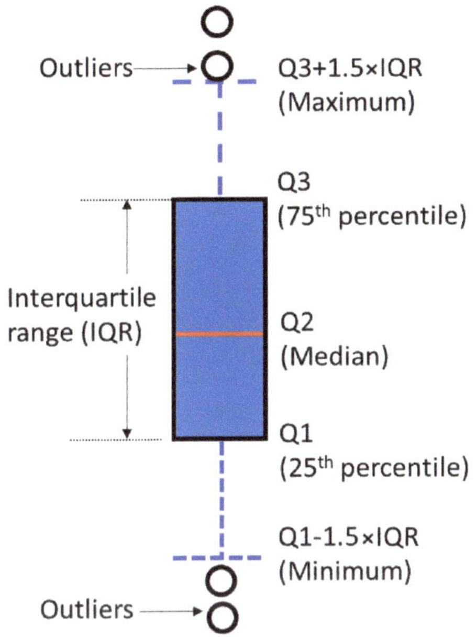

Figure 4. Key features of a box plot diagram for identifying outliers and understanding data distribution.

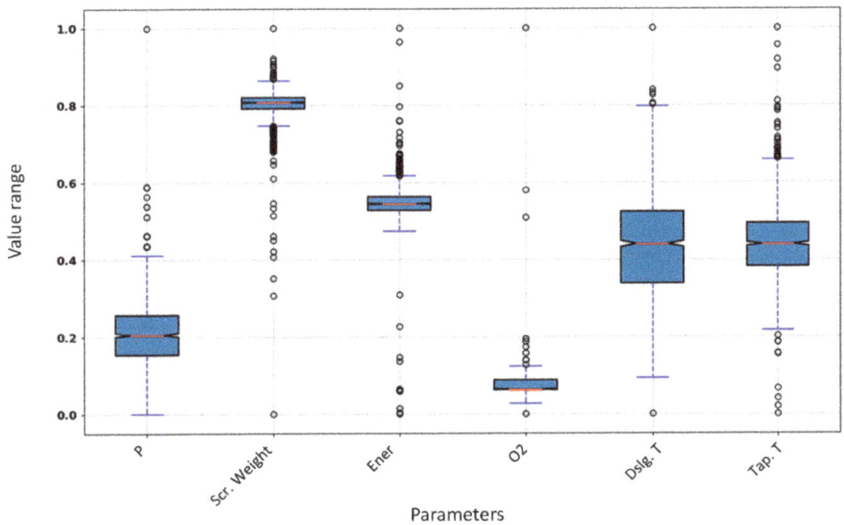

Figure 5. Descriptive statistics for the various parameters, including minimum and maximum values, mean, standard deviation, and identification of outliers for each parameter. Scr. Weight: scrap weight; Tap. T: tapping temperature; Dslg. T: deslagging temperature; Ener: energy consumption.

Following preprocessing and outlier elimination, approximately 1005 data points were retained. The number of data points is normal given the cost of preprocessing and collection. Additionally, collected over one year of plant operation, the data accurately reflect real-world conditions, making them representative of the problem at hand. The descriptive statistics for all input and output variables of the prediction models are presented in Table 2.

Table 2. Statistics describing the input and output variables for prediction.

Feature Category	Feature	Min Value	Max Value	Mean	STD *
Endpoint	P content in steel (wt%)	0.003	0.018	0.010	0.003
Scrap key composition	C content in scrap (wt%)	0.06	0.34	0.27	0.05
	Mn content in scrap (wt%)	0.58	3.58	0.80	0.10
	Cr content in scrap (wt%)	0.11	1.88	0.75	0.26
	Si content in scrap (wt%)	0.13	0.79	0.23	0.04
	S content in scrap (wt%)	0.004	0.080	0.013	0.003
Process parameters	Injected oxygen (m^3)	77.87	289.97	179.05	29.96
	Injected lime (kg)	975	1950	1048	256
	Energy consumption (kWh)	18,008	23,398	20,702	941
	Deslagging temperature (°C)	1518	1682	1600	55
	Tapping temperature (°C)	1609	1696	1652	27
	Scrap weight (kg)	41,340	43,708	42,673	783
	Process duration (min)	103	711	143	45

* STD: standard deviation.

3.2.3. Correlation Analysis and Normalization

Correlation Analysis

Correlation analysis is used to understand the relationships between independent variables and the target variable, such as the final phosphorus content in steel. This analysis clarifies the strength of associations between features and phosphorus outcomes, which is particularly valuable for ML models with simpler structures, such as RF and SVM. These models benefit from correlation insights to assess feature importance and guide feature selection, improving model interpretability, reducing training time, and enhancing learning accuracy [81]. However, for more complex models like ANNs, Moosavi-Khoonsari et al. [50] found that correlation analysis can be redundant and may even lead to decreased accuracy as ANNs can automatically learn and capture intricate, non-linear relationships. Nevertheless, exploring correlations can still offer valuable insights into feature dynamics and relationships, enhancing our understanding of the data and model behavior.

The Pearson correlation coefficients (r) and p-values were utilized for correlation analysis in this study. The r-value and t-statistic (t), used to compute p-values, are described in Equations (11) and (12), respectively. The r-values for the identified variables are depicted in Figure 6, which illustrates the linear relationships between these variables and the final phosphorus content in steel. The visualization ranks variables by the strength of their correlations, highlighting both positive and negative relationships. Values close to 1 or -1 indicate strong linear relationships, while those near 0 suggest weaker ones. Positive r-values indicate that as one variable increases, the other also increases, while negative r-values reflect an inverse relationship.

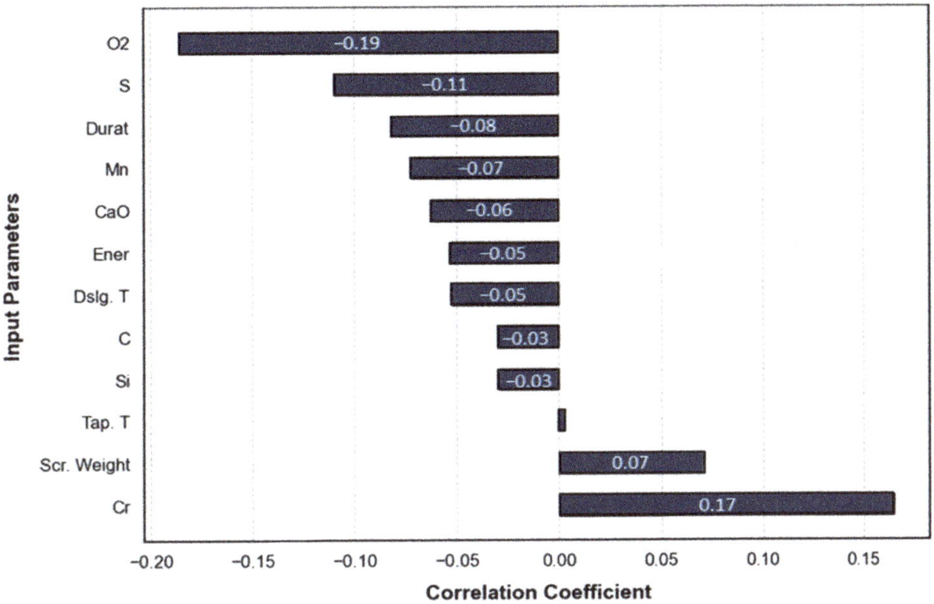

Figure 6. Pearson correlation coefficients (r) between final phosphorus content of steel (the target variable) and input parameters in electric arc furnace. Durat.: process duration; Scr. Weight: scrap weight; Tap. T: tapping temperature; Dslg. T: deslagging temperature; Ener: energy consumption.

Specifically, the analysis reveals that oxygen (O_2), sulfur (S), process duration (Durat), manganese (Mn) content of scrap, injected lime (CaO), energy consumption, deslagging temperature, and the carbon (C) and silicon (Si) contents of scrap exhibit a negative correlation with phosphorus content. Among these, the negative correlation is strongest for oxygen and weakest for carbon and silicon contents. Increasing these variables generally leads to a reduction in the phosphorus content of steel. Conversely, chromium (Cr) content of scrap, scrap weight, and tapping temperature show a positive correlation with phosphorus content, with Cr having the strongest and tapping temperature the weakest positive correlation. Increases in these variables generally result in higher final phosphorus content of steel.

$$r = \frac{\sum_{i=1}^{n} (x_i - \overline{x})(y_i - \overline{y})}{\sqrt{\sum_{i=1}^{n} (x_i - \overline{x})^2 \sum_{i=1}^{n} (y_i - \overline{y})^2}} \tag{11}$$

Let $\overline{x}$ represent the mean of the variable x; $\overline{y}$ represent the mean of the variable y; x_i denote the ith value of variable x; and y_i denote the ith value of variable y.

$$t = \frac{r\sqrt{n-2}}{\sqrt{1-r^2}} \tag{12}$$

Here, r represents the correlation coefficient; n denotes the sample size; and $n-2$ indicates the degree of freedom.

The p-values were analyzed to evaluate whether the correlations between the final phosphorus content in steel and the input parameters, as listed in Table 3, were statistically significant. A p-value less than 0.01 indicates that the correlation is very significant, suggesting a strong likelihood that the observed relationship is not due to chance. A p-value less than 0.05 signifies that the correlation is significant, though less robust than those with p-values below 0.01. Conversely, a p-value greater than 0.05 implies that the correlation is not statistically significant, indicating that the relationship may be due to random variation [82]. Based on the p-values, the relationships between the input parameters and the final phosphorus content in steel can be categorized into three levels of significance. The most statistically significant correlations, with p-values less than 0.01, include injected oxygen ($p = 3 \times 10^{-9}$), Cr content in scrap ($p = 1.39 \times 10^{-7}$), S content in scrap ($p = 5 \times 10^{-4}$), and the process duration ($p = 8.89 \times 10^{-3}$). These parameters exhibit strong relationships with phosphorus content, indicating highly significant correlations. In the intermediate category, with p-values between 0.01 and 0.05, are scrap weight ($p = 2.44 \times 10^{-2}$), Mn content in scrap ($p = 2.22 \times 10^{-2}$), and injected lime ($p = 4.73 \times 10^{-2}$), suggesting these variables have notable but less pronounced effects. Finally, parameters with p-values greater than 0.05, including energy consumption ($p = 9.23 \times 10^{-2}$), deslagging temperature ($p = 9.35 \times 10^{-2}$), C content in scrap ($p = 3.49 \times 10^{-1}$), Si content in scrap ($p = 3.56 \times 10^{-1}$), and tapping temperature ($p = 9.38 \times 10^{-1}$), show weaker and statistically insignificant correlations with the final phosphorus content of steel.

Table 3. Calculated results of p-value between final phosphorous content of steel and input variables.

Input Parameters	r	p**-Value**
Scrap weight	0.07	2.44×10^{-2} *
C content in scrap (kg)	−0.03	3.49×10^{-1}
Mn content in scrap (kg)	−0.07	2.22×10^{-2} *
Cr content in scrap (kg)	0.17	1.39×10^{-7} **
Si content in scrap (kg)	−0.03	3.56×10^{-1}
S content in scrap (kg)	−0.11	5×10^{-4} **
Injected oxygen (kg)	−0.18	3×10^{-9} **
Injected lime (kg)	−0.06	4.73×10^{-2} *
Energy consumption	−0.05	9.23×10^{-2}
Deslagging temperature	−0.05	9.35×10^{-2}
Tapping temperature	0.005	9.38×10^{-1}
Process duration	−0.08	8.89×10^{-3} *

A total of 1005 data points were analyzed. p-values < 0.05 are marked with (*), and p-values < 0.01 with (**).

There is a direct correlation between the injected O_2 and the final phosphorus content in steel. The injection of O_2 promotes the oxidation of scrap, increasing the FeO content in the slag. As detailed in Section 2.3, the presence of FeO and CaO enhances the phosphorus removal process [19–22,63,64]. Increasing the amount of added CaO decreases phosphorus content in steel by raising the slag's basicity. Conversely, an increase in chromium (Cr) leads to an increase in the final P content of the steel. Karbowniczek et al. [75] also demonstrated that increasing the chromium content in metal and the Cr_2O_3 content in slag results in a decrease in phosphorus removal capacity, irrespective of other parameters. Chromium in scrap oxidizes to Cr_2O_3 in the slag, which diminishes the dephosphorization capacity because Cr_2O_3 is an acidic component that reduces slag basicity. Additionally, chromium stabilizes phosphorus in the metal phase, as indicated by the first-order interaction coefficient ($e_P^{Cr} = -0.93$ [79], -0.03 [76]). Moreover, Cr_2O_3 promotes the formation of spinel solid particles, which reduces the proportion of liquid slag and increases its viscosity. This, in turn, lowers the dephosphorization capacity by decreasing the amount of liquid slag available for phosphorus removal and potentially slowing the kinetics of dephosphorization. Yang et al. [77] also reported that Cr_2O_3 levels exceeding 0.5 wt% in slag lead to increased viscosity. An increase in sulfur (S) content in scrap results in a decrease in phosphorus content in steel. This effect can be attributed to the increased activity coefficient of phosphorus in the presence of sulfur in the metal ($e_P^S = 0.028$ [76], 0.048 [79]). Prolonging the process duration also aids dephosphorization by increasing the time for phosphorus partitioning to the slag via the steel–slag interface. The oxidation of manganese (Mn) from the scrap leads to a high MnO content in the slag. MnO increases slag basicity, decreases viscosity, and lowers the liquidus temperature [78]. Consequently, it is expected to enhance the dephosphorization process, as observed in this study. An increase in scrap weight raises the phosphorus content in steel, as scrap serves as a source of phosphorus in the process.

Data Normalization

In ML models, various parameters with distinct values and units are utilized. To facilitate the learning process and ensure rapid model convergence while mitigating bias from differing scales, it is crucial to standardize the data. The Min–Max normalization method is used for this purpose, scaling values to the range [0, 1]. The normalization process is mathematically expressed in Equation (13):

$$z_i = \frac{x_i - \min(x)}{\max(x) - \min(x)} \tag{13}$$

where z_i is the normalized value, x_i is the original value, $\max(x)$ denotes the maximum value of the data, and $\min(x)$ denotes the minimum value of the data.

3.3. Model Evaluation

The efficiency of the ML models was evaluated using several statistical metrics, including *MSE*, *RMSE*, coefficient of determination (R^2), and r. The mathematical formulas for calculating *MSE*, *RMSE*, and R^2 are provided in Equations (14)–(16).

$$MSE = \frac{1}{N}\sum_{j=1}^{N}\left(y_j - \hat{y}_j\right)^2 \tag{14}$$

$$RMSE = \sqrt{\frac{1}{N}\sum_{j=1}^{N}\left(y_j - \hat{y}_j\right)^2} \tag{15}$$

$$R^2 = 1 - \frac{\sum_{i=1}^{m}\left(\hat{y}_j - y_i\right)^2}{\sum_{i=1}^{m}(\overline{y} - y_i)^2} \tag{16}$$

where N is the number of the entire dataset; y_j is the ith actual value of y; $\hat{y}_j$ is the ith predicted value of $\hat{y}$; and $\overline{y}$ is the average of the actual values.

For regression models, the correlation coefficient r can also be used to evaluate the relationship between predicted values and actual values (refer to Equation (11)). However, it is often supplemented with other metrics such as R^2, *MSE*, or *RMSE* for a more comprehensive evaluation.

4. Results and Discussion

4.1. Hyperparameter Optimization of ANN

As mentioned in the Section "Establishment of Artificial Neural Network Models", the hyperparameters were optimized using the approach proposed by Begstra et al. [74,83]

to develop an optimal ANN model. The goal was to design architectures tailored to the specific problem at hand. Various combinations were tested, with different numbers of hidden layers, neurons, and iterations, while considering the required learning time. Table 4 provides a summary of the ANN structures tested. After an extensive series of trials and evaluations, a configuration that was both simple and effective was identified. The Adam optimizer was used to manage the learning rate, facilitating faster model convergence.

Table 4. Hyperparameters used in the developed ANN models.

Hyperparameters	Different Models		
	ANN (1) 12-16-8-1	**ANN (2) 12-144-256-64-1**	**ANN (3) 12-128-128-128-64-1**
Number of neurons	24	464	448
Number of layers	2	3	4
Number of epochs	1000	100	500
Batch size	50	50	50

The first tested ANN model (ANN (1)) featured two hidden layers, with sixteen neurons in the first layer and eight neurons in the second layer, and was trained for 50,000 iterations, as shown in Figure 7a. The figure displays two curves: the training loss (in blue) and the validation loss (in orange), plotted against the total number of iterations. Both curves exhibit a similar pattern of gradual decline, suggesting a reduction in loss over time. After 50,000 iterations, the model achieved an *MSE* value of 0.0148.

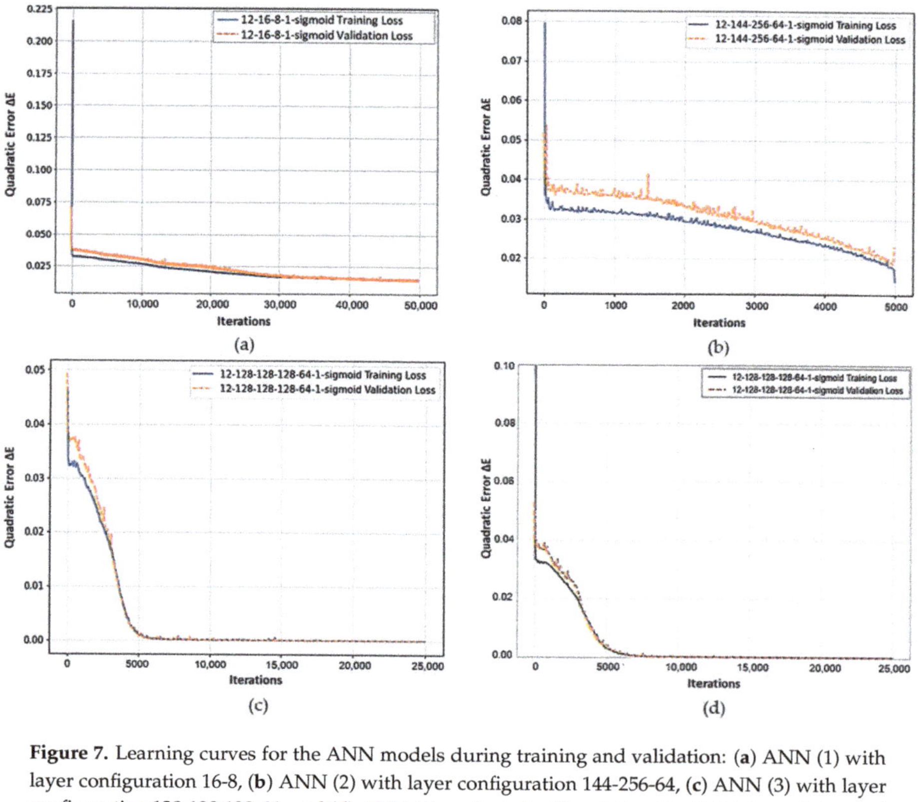

Figure 7. Learning curves for the ANN models during training and validation: (**a**) ANN (1) with layer configuration 16-8, (**b**) ANN (2) with layer configuration 144-256-64, (**c**) ANN (3) with layer configuration 128-128-128-64, and (**d**) ANN (3) evaluated with a data split of 80% for training and 20% for testing.

A second architectural configuration (ANN (2)) was tested, featuring a more complex model with additional hidden layers and neurons compared to the previous one. This configuration included 144 neurons in the first hidden layer, 256 in the second, and 64 in the third, with a total of 5000 iterations. As shown in Figure 7b, the number of iterations was insufficient, as the convergence curves did not stabilize. The minimum *MSE* achieved with this configuration was 0.0097.

The third architectural configuration (ANN (3)) was tested to improve the model's precision and overall accuracy. This iteration introduced an additional hidden layer, resulting in a total of four hidden layers, making the model moderately more complex than the previous two. In this configuration, the number of neurons per layer was reduced, with 128 neurons in the first three hidden layers and 64 in the fourth layer, and the model was trained for 25,000 iterations. As shown in Figure 7c, the *MSE* approached zero after approximately 6000 iterations, with a minimum *MSE* of 0.00003.

As a concluding phase in the process of optimizing the ANN model, the model ANN (3), which has the following architectural parameters, 128-128-128-64, was selected. A minor alteration to the dataset was implemented by combining the validation set with the training set, thus enabling the model to be retrained with a more extensive dataset. The initial division of the total dataset into three subsets was as follows: 60% for training, 20% for validation, and 20% for testing. In the final stage of the optimization process, the dataset was divided into two subsets: 80% for training and 20% for testing. This approach ensures that the model demonstrates generalization capabilities that extend well beyond the parameters of the training data. At last, the model was evaluated based on its performance when tested with the test set. The results of the model convergence are presented in Figure 7d. It is evident that the model converges at a faster rate than previous models, reaching a minimum *MSE* value after just 5000 iterations. For the model trained on a CPU (Intel(R) Core(TM) i7-4710HQ CPU @ 2.50 GHz) and GPU (Intel(R) HD Graphics 4600 with a memory capacity of 12.0 GB), the training process took approximately 90 min.

4.2. Comparison of the ANN Models with Other Models

To evaluate the generalization performance and prediction accuracy of the models developed in this study, various metrics were used. Table 5 summarizes the results of the performance evaluation for all the tested models, including ANN models with varying hyperparameters, the RF model, and the SVM-RBF model.

Table 5. Metrics for the developed machine learning models.

Metrics	Model					
	SVM-RBF	**RF**	**ANN (1)**	**ANN (2)**	**ANN (3)**	**ANN (3) Optimized**
MSE	0.03	7.9034×10^{-6}	0.0148	0.0097	0.00003	0.000016
RMSE	0.17	0.0028	0.1216	0.0985	0.0055	0.004999
r	0.2828	0.3316	0.778	0.866	0.9996	0.9998
R^2 *	0.08	0.11	0.61	0.75	0.9993	0.9996

* R^2 value for the test set.

Based on the r and p-values, six parameters with weak correlations were eliminated. The updated results for the modified RF model, which was evaluated using a dataset split into 80% for training (with 5-fold cross-validation) and 20% for testing, are as follows: the best R^2 score during cross-validation is 0.1236, while the R^2 scores on the validation and test sets are 0.12 and 0.11, respectively. The training score is notably higher at 0.56, indicating that despite the parameter adjustments, the model still exhibits a limited ability to generalize to unseen data, with potential overfitting observed.

For comparison, the RF model with 12 input parameters was also evaluated across different datasets, which were divided into three parts: 60% for training, 20% for validation, and 20% for testing. The R^2 value on the validation set was 0.10, while the R^2 value on the test set was 0.07. In contrast, the training score (R^2) was 0.87. These metrics suggest that the model's generalization performance improved only slightly, with an R^2 value of 0.11 for the six input parameters compared to 0.07 for the twelve input parameters.

The *MSE* and *RMSE* for the SVM-RBF model are 0.03 and 0.17, respectively. The R^2 values for training, validation, and testing are 0.17, 0.09, and 0.08, respectively. The low R^2 values indicate that the SVM model with an RBF kernel does not capture the relationship between the input variables and the target variable well. This suggests that the model is not performing effectively. An R^2 value of 0.08 for the test set means that the model explains only 8% of the variance in the test data, confirming that the model does not generalize well to new data. These results suggest that neither the RF nor the SVM-RBF models effectively captured the correlation between the input variables and the target, indicating that these models may not be well suited for this problem.

The ANN (1) model achieved an *MSE* of 0.0148, *RMSE* of 0.1216, R^2 of 0.61, and r of 0.778. The ANN (2) model showed improvement with an *MSE* of 0.0097, *RMSE* of 0.0985, R^2 of 0.75, and r of 0.866. The ANN (3) model, with its architecture modifications, demonstrated outstanding results: *MSE* of 0.00003, *RMSE* of 0.0055, R^2 of 0.9993, and r of 0.9996. These results indicate that this architecture is highly effective for explaining the relationship between input variables of the EAF and the endpoint phosphorous content of steel. Finally, the optimized ANN (3) model, which involved dataset adjustments but retained the same architecture, achieved the following metrics: *MSE* of 0.000016, *RMSE* of 0.00499, R^2 of 0.9996, and r of 0.9998.

To validate the ANN models, regression plots were used, as shown in Figure 8, to illustrate the relationship between the ANN model's outputs and the actual values. As seen, the precision of phosphorus measurement is ±0.001 wt%, with 15 distinct phosphorus concentration values across all samples. A comparison of the scatterplots for the three ANN models shows that ANN (3)'s data distribution is significantly closer to the dashed line, with a high degree of overlap, compared to the other models.

Table 6 compares the performance metrics of the ANN models developed in this work with those of previous models in the literature, summarizing various studies that focus on different modeling approaches for predicting endpoint phosphorus content in steel during EAF and BOF processes. It details the models employed, input parameters, dataset sizes, and evaluation metrics. The ANN (3) model, which used 12 input parameters, analyzed a dataset of 1763 data points (with 1005 utilized), achieving an R^2 of 0.9996 and an r of 0.9998. In contrast, the ANN (2) model, which was also applied to the same dataset, yielded an R^2 of 0.75 and r of 0.866. Various models developed by Zhang et al. [46] revealed moderate performance, with the highest r being 0.608 for RF and the lowest at 0.382 for ridge regression. Both BPNN models by Zhou et al. [52] showed R^2 values of 0.7596 and 0.8456, indicating decent predictive capabilities. The unhybrid ANN and hybrid physics-based ANN developed by Wang et al. [48] showed NRMSE values of 0.1796 and 0.1775, respectively. Chang et al. [43] developed various models and the R^2 values ranged from 0.280 for FCN to 0.729 for the multi-channel GCN. He and Zhang [41] achieved an r-value of 0.79 for PCA-BPNN. Laha et al. [40] also developed various models for a reverberatory furnace, achieving an R^2 of 82% for the SVR model.

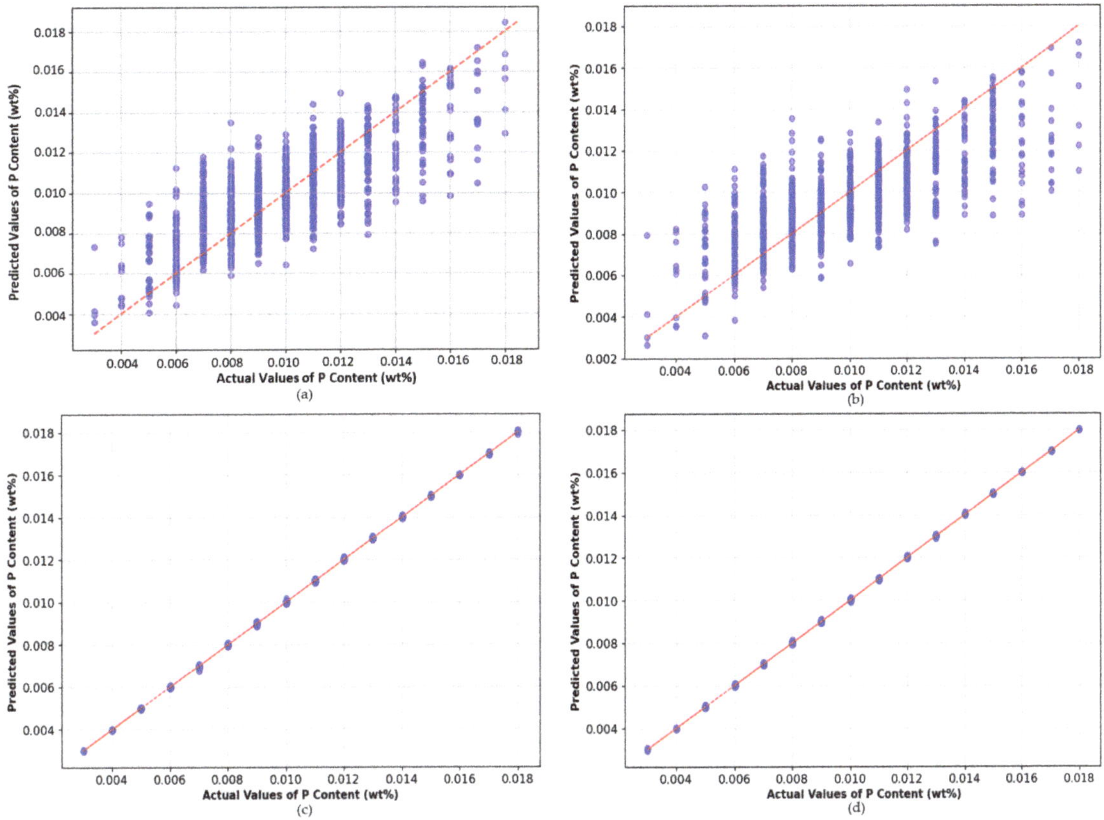

Figure 8. Comparison between predicted P values obtained by the ANN models and actual measured values: (**a**) ANN (1): 16-8; (**b**) ANN (2): 144-256-64; (**c**) ANN (3): 128-128-128-64; and (**d**) ANN (3) evaluated with a data split of 80% for training and 20% for testing.

Table 6. Comparative summary of the newly proposed model and previous models.

References	Process	Model	Input Parameters	Data Size	Evaluation Metrics
This work	EAF (Scrap)	ANN (3) ANN (2)	12	1763 (1005)	R^2: 0.9996 r: 0.9998 R^2: 0.75 r: 0.866
Zou et al. [47]	EAF (HM * + Scrap)	BPNN	10	1250 (580)	- -
Chen et al. [44]	EAF (HM + Scrap)	k means-BPNN-DT DNN BPNN	18	1258 (1114)	- - -
Yuan et al. [34]	EAF	LS-SVM-PCR	10	82	-

Table 6. *Cont.*

References	Process	Model	Input Parameters	Data Size	Evaluation Metrics
Zhang et al. [46]	BOF	Ridge regression	16	13,000 (7776)	*r*: 0.382 MARE: 0.182 *RMSE*: 0.00369
		GBR			*r*: 0.599 MARE: 0.155 *RMSE*: 0.00325
		SVM			*r*: 0.52 MARE: 0.177 *RMSE*: 0.00342
		RF			*r*: 0.608 MARE: 0.156 *RMSE*: 0.00319
		CNN			*r*: 0.541 MARE: 0.173 *RMSE*: 0.00354
Zhou et al. [52]	BOF	Unconstrained BPNN	10	(900)	R^2: 0.7596 *RMSE*: 0.0037
		Monotone-constrained BPNN			R^2: 0.8456 *RMSE*: 0.0030
Wang et al. [48]	BOF	Unhybrid NN	19	28,000	NRMSE: 0.1796
		Hybrid physics-based NN			NRMSE: 0.1775
Chang et al. [43]	BOF	PLS	42		R^2: 0.728 *RMSE*: 0.0019
		SVR			R^2: 0.622 *RMSE*: 0.0022
		FCN			R^2: 0.280 *RMSE*: 0.0028
		ELM			R^2: 0.620 *RMSE*: 0.0022
		GCN			R^2: -0.132 *RMSE*: 0.0038
		Multi-channel GCN			R^2: 0.729 *RMSE*: 0.0019
He and Zhang [41]	BOF	PCA and BPNN	18 (7 with PCA)	1978	*r*: 0.79
Laha et al. [40]	Reverberatory Furnace	RF, NN, DENFIS, SVR	10	54	R^2: 82% (for SVR)

* HM stands for hot metal.

The hit rate measures the percentage of predictions within a specified error margin for the final phosphorus content in steel. It evaluates the model's generalization ability using unseen data, which helps indicate the risk of overfitting. Figure 9 illustrates the hit rates of the three ANN models developed in this work compared with those of models from the literature. The calculated minimum range for phosphorus variation is ±0.001%, which is reasonable given the precision of the phosphorus measurements. For ANN (2), hit rates were 45% within ±0.001 wt% (10 ppm P), 72% within ±0.002 wt% (20 ppm P), 87% within ±0.003 wt% (30 ppm P), and 95% within ±0.004 wt% (40 ppm P). However, both ANN (3) and the optimized ANN (3) with an 80%–20% split achieved a hit rate of 100% across all error thresholds.

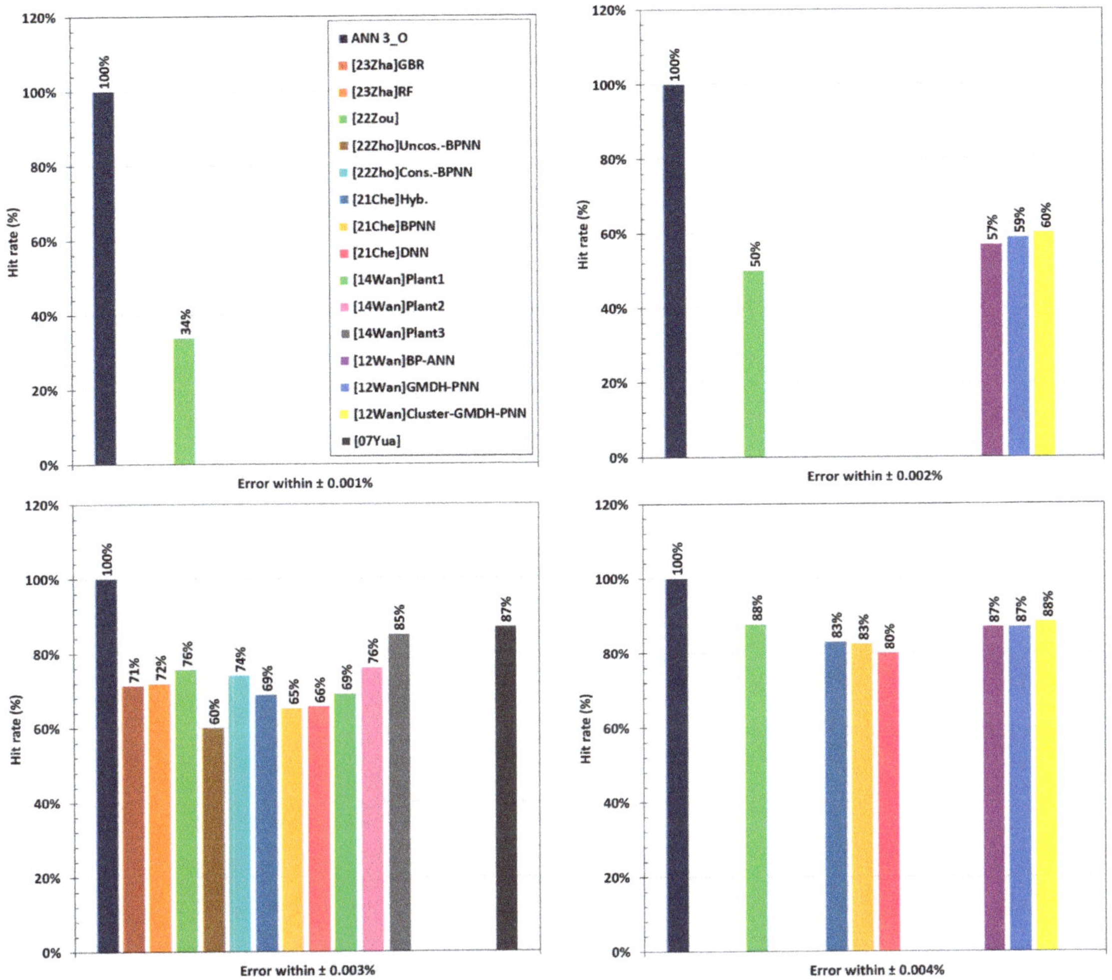

Figure 9. Hit rates of the optimized model ANN 3_O (80%–20% split) along with those of previously developed models [34,44,46,47,52,67].

Both ANN (2) and ANN (3) outperform earlier models [34,44,46,47,52], with ANN (3) showing particular accuracy in predicting the final phosphorus content in steel. The predictive accuracy of different NN architectures and their derivatives varies based on the input parameters and model design [41,43,44,46–48,50,52]. It might be that including scrap composition in our model appears to have enhanced prediction accuracy, as evidenced by a relatively strong correlation between scrap composition and final phosphorus content in steel (see Figure 5). Additionally, compared to previous work, neither a DNN with seven hidden layers and 416 neurons nor an ANN with two hidden layers and 24 neurons [50] could match the accuracy of the ANN (3) model with four hidden layers and 448 neurons. Additionally, ANN (2), despite its more complex nature, performs less well than ANN (3). This could be due to the fact that having four hidden layers in ANN (3) instead of three, as in ANN (2), enables the model to capture more complex patterns in the data. Furthermore, the ANN model with more neurons (464 in ANN (2)) may be more prone to overfitting, whereas ANN (3), with fewer neurons (448), could be less complex and better able to generalize. When searching a grid of neurons (y-axis) and hidden layers (x-axis), ranging from two to seven hidden layers and 24 to 464 neurons, as tested in both this work and previous work [50], we find that the combination of four hidden layers and 448 neurons provides the best fit for this specific dataset, leading to improved performance. This observation underscores the importance of carefully selecting input parameters, data processing methods, and designing the architecture of NN models.

By creating a user-friendly interface, the ANN (3) model can be effectively used for practical implementation, allowing plant engineers and operators to utilize it in situ for optimization purposes. The effect of different process parameters and initial input data on the endpoint phosphorous content can be predicted. Additionally, the proposed model architecture can be adapted to different scrap-based EAF operations with some adjustments.

The proposed model has the potential for scalability, particularly as a generic pre-trained model for scrap-based EAF steelmaking. This model can be tested against larger datasets and continuously improved and optimized to adapt to the dynamic conditions of steelmaking. Such adaptability is intrinsic to any robust machine learning model, ensuring its relevance as operational parameters evolve. Further development and customization of the model for specific furnaces or plants is encouraged; however, its current use remains subject to intellectual property constraints, which require verification and approval from industrial partners.

The current model can be incorporated into existing metallurgical workflows by developing an accessible interface for plant users and engineers. This interface would allow them to predict the impact of initial conditions and operational parameters on steel phosphorus content and quality. With the anticipated decline in scrap quality, having a model that considers variations in scrap composition and predicts its effects on phosphorus content and steel quality is critical for operational decision-making. The model can guide scrap selection and blending and also assist in evaluating and adjusting various operational parameters to compensate for shifts in scrap composition, making it a valuable tool for maintaining efficiency and product quality in the evolving landscape of steelmaking.

5. Conclusions

In the present work, various machine learning models were used to predict the phosphorus content in a low-alloy steel at the end of a scrap-based electric arc furnace (EAF) process. The tested models included a support vector machine (SVM) with a radial basis function (RBF) kernel, a random forest (RF), and artificial neural networks (ANNs). The main findings of this research are summarized below.

- Strong correlations with the endpoint phosphorus content of steel were found for Cr and S contents in scrap, injected oxygen, and process duration (p-value < 0.01). Intermediate correlations were observed for scrap weight, Mn content in scrap, and injected lime (0.01 < p-value < 0.05). Weaker correlations were noted for energy consumption, deslagging temperature, C and Si contents in scrap, and tapping temperature (p-value > 0.05).
- Machine learning models, such as SVM-RBF and RF, did not yield satisfactory results in terms of phosphorus prediction accuracy. Several ANN models with different architectures were tested, and the best model consisted of four hidden layers and 448 neurons. This model was trained for 500 epochs with batches of 50 samples, and implemented using the TensorFlow library. Hyperparameters were carefully tuned to maximize performance, employing the Adam optimizer for adaptive learning rate adjustments and the sigmoid activation function to introduce non-linearity in each neuron.

- The optimized ANN model achieved higher performance compared to similar models reported in the literature, with a root mean square error (*RMSE*) of 0.004999, a mean squared error (*MSE*) of 0.000016, a correlation coefficient (r) of 0.9998, and a coefficient of determination (R^2) of 0.9996. Additionally, it demonstrated a very good hit rate of 100% for predicting endpoint phosphorus content within ±0.001 wt% in steel (when tested on over 200 unseen data points). These results confirm that, even with a limited dataset (1005), an optimized ANN architecture combined with proper input data selection, such as scrap composition, can deliver accurate and reliable predictions of the phosphorus content in steel during the EAF process.

Author Contributions: Conceptualization, R.A. and E.M.-K.; Methodology, R.A. and S.E.K.; Software, R.A.; Validation, R.A., S.E.K. and E.M.-K.; Formal analysis, R.A.; Resources, E.M.-K. and M.J.; Data curation, R.A.; Writing—original draft, R.A., E.M.-K.; Writing—review & editing, M.J., S.E.K. and E.M.-K.; Visualization, R.A., E.M.-K.; Supervision, M.J., S.E.K. and E.M.-K.; Project administration, E.M.-K.; Funding acquisition, M.J. and E.M.-K. All authors have read and agreed to the published version of the manuscript.

Funding: This research was funded by Finkl Steel-Sorel and Mitacs Accelerate Program (IT28458).

Data Availability Statement: The original contributions presented in the study are included in the article, further inquiries can be directed to the corresponding author.

Acknowledgments: In addition, the authors highly appreciate Finkl Steel-Sorel for providing the plant data and technical discussion throughout the project. The authors also thank CIFAR for their kind contribution and support of the project.

Conflicts of Interest: The authors declare that this study received funding from Finkl Steel-Sorel. The funder was not involved in the study design, collection, analysis, interpretation of data, the writing of this article or the decision to submit it for publication.

Appendix A

The libraries utilized, along with their respective versions, for developing the ANN models are listed below:

Core Libraries
numPy 2.0.2
pandas 2.2.3
scipy 1.14.1
scikit-learn 1.5.2
Visualization
matplotlib 3.9.2
contourpy 1.3.0
cycler 0.12.1
fonttools 4.54.1
kiwisolver 1.4.7
pillow 11.0.0
Pygments 2.18.0
rich 13.9.4
Machine Learning/Deep Learning
tensorflow 2.18.0
tensorflow_intel 2.18.0
keras 3.6.0
tensorboard 2.18.0
tensorboard-data-server 0.7.2

tensorflow-io-gcs-filesystem 0.31.0
opt_einsum 3.4.0
gast 0.6.0
grpcio 1.67.1
h5py 3.12.1
AI/Language Models
openai 1.51.2
Data Manipulation and Parsing
openpyxl 3.1.5
et_xmlfile 2.0.0
python-dateutil 2.9.0.post0
pytz 2024.2
attrs 24.2.0
packaging 24.1
markdown-it-py 3.0.0
mdurl 0.1.2
Web and Networking
requests 2.32.3
httpcore 1.0.6
httpx 0.27.2
urllib3 2.2.3
websocket-client 1.8.0
websockets 13.1
sniffio 1.3.1
wsproto 1.2.0
Serialization and Protocol Buffers
protobuf 5.28.3
flatbuffers 24.3.25
Automation and GUI Tools
PyAutoGUI 0.9.54
PyGetWindow 0.0.9
PyMsgBox 1.0.9
PyRect 0.2.0
PyScreeze 1.0.1
Pyperclip 1.9.0
Pytweening 1.2.0
MouseInfo 0.1.3
PyInstaller and Related
Pyinstaller 6.11.1
pyinstaller-hooks-contrib 2024.10
Database Connectivity
mysql-connector-python 9.1.0
PyQt
PyQt5 5.15.11
PyQt5-Qt5 5.15.2
PyQt5_sip 12.15.0
Miscellaneous Utilities
tqdm 4.66.5
colorama 0.4.6
namex 0.0.8

distro 1.9.0
termcolor 2.5.0
Werkzeug 3.1.2
wrapt 1.16.0
Typing and Validation
pydantic 2.9.2
pydantic_core 2.23.4
typing_extensions 4.12.2
annotated-types 0.7.0
Selenium and Browsers
selenium 4.25.0
undetected-chromedriver 3.5.5
Async Libraries
anyio 4.6.2.post1
trio 0.27.0
trio-websocket 0.11.1
outcome 1.3.0.post0
Additional Libraries
absl-py 2.1.0
altgraph 0.17.4
astunparse 1.6.3
certify 2024.8.30
cffi 1.17.1
charset-normalizer 3.4.0
exceptiongroup 1.2.2
h11 0.14.0
joblib 1.4.2
Markdown 3.7
ml-dtypes 0.4.1
optree 0.13.0
pefile 2023.2.7
six 1.16.0
sortedcontainers 2.4.0
tzdata 2024.2
idna 3.10
pycparser 2.22
pywin32-ctypes 0.2.3
jiter 0.6.1

References

1. Abadi, M.M.; Tang, H.; Rashidi, M.M. A Review of Simulation and Numerical Modeling of Electric Arc Furnace (EAF) and its Processes. *Heliyon* **2024**, *10*, e32157. [CrossRef]
2. Kildahl, H.; Wang, L.; Tong, L.; Ding, Y. Cost effective decarbonisation of blast furnace–basic oxygen furnace steel production through thermochemical sector coupling. *J. Clean. Prod.* **2023**, *389*, 135963. [CrossRef]
3. WorldSteelAssociation. Maximising Scrap Use Helps Reduce CO_2 Emissions. Raw Materials 2024. Available online: https://worldsteel.org/steel-topics/raw-materials/ (accessed on 1 September 2024).
4. Heo, J.H.; Park, J.H. Effect of Slag Composition on Dephosphorization and Foamability in the Electric Arc Furnace Steelmaking Process: Improvement of Plant Operation. *Metall. Mater. Trans. B* **2021**, *52*, 3613–3623. [CrossRef]
5. Lin, W.; Jiao, S.; Zhou, K.; Sun, J.; Feng, X.; Liu, Q. A review of multi-phase slag refining for dephosphorization in the steelmaking process. *Front. Mater.* **2020**, *7*, 602522. [CrossRef]

6. Rodrigues, C.; Bandeira, R.; Duarte, B.; Tremiliosi-Filho, G.; Jorge, A.M., Jr. Effect of phosphorus content on the mechanical, microstructure and corrosion properties of supermartensitic stainless steel. *Mater. Sci. Eng. A* **2016**, *650*, 75–83. [CrossRef]
7. Holappa, L.; Nava, A.C. Secondary steelmaking. In *Treatise on Process Metallurgy*; Elsevier: Amsterdam, The Netherlands, 2024; pp. 267–301.
8. Menard, P. Finkl Steel Sorel, Saint-Joseph-de-Sorel, QC, Canada. Personal Communication, 2023.
9. Compañero, R.J.; Feldmann, A.; Tilliander, A. Circular steel: How information and actor incentives impact the recyclability of scrap. *J. Sustain. Metall.* **2021**, *7*, 1654–1670. [CrossRef]
10. Anameric, B.; Rohaus, D.; Riebeiro, T.R. Ironmaking. In *SME Mineral Processing & Extractive Metallurgy Handbook*; Dunne, R.C., Kawatra, S.K., Young, C.A., Eds.; Society for Mining, Metallurgy, and Exploration (SME): Englewood, CO, USA, 2019; pp. 1781–1796.
11. Ripke, S.J.; Poveromo, J.; Battle, T.P.; Al, E. Iron ore beneficiation. In *SME Mineral Processing & Extractive Metallurgy Handbook*; Dunne, R.C., Kawatra, S.K., Young, C.A., Eds.; Society for Mining, Metallurgy, and Exploration (SME): Englewood, CO, USA, 2019; pp. 1755–1779.
12. Suito, H.; Inoue, R.; Takada, M. Phosphorus distribution between liquid iron and MgO saturated slags of the system CaO-MgO-FeOx-SiO_2. *Tetsu-to-Hagané* **1981**, *67*, 2645–2654. [CrossRef]
13. Suito, H.; Inoue, R. Effect of calcium fluoride on phosphorus distribution between MgO-saturated slags of the system CaO-MgO-FeOx-SiO_2 and liquid iron. *Tetsu-to-Hagané* **1982**, *68*, 1541–1550. [CrossRef] [PubMed]
14. Suito, H.; Inoue, R. Effects of Na_2O and BaO additions on phosphorus distribution between CaO-MgO-FetO-SiO_2-slags and liquid iron. *Trans. Iron Steel Inst. Jpn.* **1984**, *24*, 47–53. [CrossRef]
15. Nakamura, S.; Tsukihashi, F.; Sano, N. Phosphorus partition between CaOsatd.-BaO-SiO_2-FetO slags and liquid iron at 1873 K. *ISIJ Int.* **1993**, *33*, 53–58. [CrossRef]
16. Ostrovski, O.I.; Utochkin, Y.I.; Pavlov, A.V.; Akberdin, R.A. Phosphate Capacity of the CaO-CaF_2 System Containing Chromium Oxide. *ISIJ Int.* **1994**, *34*, 849–851. [CrossRef]
17. Im, J.; Morita, K.; Sano, N. Phosphorus distribution ratios between CaO-SiO_2-FetO slags and carbon-saturated iron at 1573 K. *ISIJ Int.* **1996**, *36*, 517–521. [CrossRef]
18. Katsuki, J.-i.; Yashima, Y.; Yamauchi, T.; Hasegawa, M. Removal of P and Cr by oxidation refining of Fe-36% Ni melt. *ISIJ Int.* **1996**, *36*, S73–S76. [CrossRef] [PubMed]
19. Hamano, T.; Tsukihashi, F. The Effect of B_2O_3 on Dephosphorization of Molten Steel by FeOx-CaO-MgOsatd.-SiO_2 Slags at 1873K. *ISIJ Int.* **2005**, *45*, 159–165. [CrossRef]
20. Lee, C.; Fruehan, R. Phosphorus equilibrium between hot metal and slag. *Ironmak. Steelmak.* **2005**, *32*, 503–508. [CrossRef]
21. Li, G.; Hamano, T.; Tsukihashi, F. The effect of Na_2O and Al_2O_3 on dephosphorization of molten steel by high basicity MgO saturated CaO-FeOx-SiO_2 slag. *ISIJ Int.* **2005**, *45*, 12–18. [CrossRef]
22. Basu, S.; Lahiri, A.K.; Seetharaman, S. Phosphorus partition between liquid steel and CaO-SiO_2-P_2O_5-MgO slag containing low FeO. *Metall. Mater. Trans. B* **2007**, *38*, 357–366. [CrossRef]
23. Cho, M.K.; Park, J.H.; Min, D.J. Phosphate Capacity of CaO–SiO_2–MnO–FeO Slag Saturated with MgO. *ISIJ Int.* **2010**, *50*, 324–326. [CrossRef]
24. Li, F.; Li, X.; Yang, S.; Zhang, Y. Distribution ratios of phosphorus between CaO-FeO-SiO_2-Al_2O_3/Na_2O/TiO_2 slags and carbon-saturated iron. *Metall. Mater. Trans. B* **2017**, *48*, 2367–2378. [CrossRef]
25. Drain, P.B.; Monaghan, B.J.; Longbottom, R.J.; Chapman, M.W.; Zhang, G.; Chew, S.J. Phosphorus partition and phosphate capacity of basic oxygen steelmaking slags. *ISIJ Int.* **2018**, *58*, 1965–1971. [CrossRef]
26. Heo, J.H.; Park, J.H. Effect of direct reduced iron (DRI) on dephosphorization of molten steel by electric arc furnace slag. *Metall. Mater. Trans. B* **2018**, *49*, 3381–3389. [CrossRef]
27. Frueham, R.J. *AISI/DOE Technology Roadmap Program: Behavior of Phosphorus in DRI/HBI During Electric Furnace Steelmaking*; American Iron and Steel Institute (US): Pittsburgh, PA, USA, 2001.
28. Lee, M.; Trotter, D.; Mazzei, O. The production of low phosphorus and nitrogen steels in an EAF using HBI. *Scand. J. Metall.* **2008**, *30*, 286–291. [CrossRef]
29. Hassan, A.; Kotelnikov, G.; Semin, A.; Megahed, G. Phosphorous behavior in Electric Arc Furnace steelmaking with the melting of high phosphorous content direct reduced iron. In Proceedings of the METAL 2015-24th International Conference on Metallurgy and Materials, Conference Proceedings, Brno, Czech Republic, 3–5 June 2015.
30. Odenthal, H.J.; Kemminger, A.; Krause, F.; Sankowski, L.; Uebber, N.; Vogl, N. Review on modeling and simulation of the electric arc furnace (EAF). *Steel Res. Int.* **2018**, *89*, 1700098. [CrossRef]
31. Ek, M.; Shu, Q.; van Boggelen, J.; Sichen, D. New approach towards dynamic modelling of dephosphorisation in converter process. *Ironmak. Steelmak.* **2012**, *39*, 77–84. [CrossRef]
32. Hay, T.; Visuri, V.-V.; Aula, M.; Echterhof, T. A review of mathematical process models for the electric arc furnace process. *Steel Res. Int.* **2021**, *92*, 2000395. [CrossRef]

33. Tao, J.; Qian, W. Intelligent Method For BOF Endpoint [P]&[Mn] Estimation. In Proceedings of the 2006 6th World Congress on Intelligent Control and Automation, Dalian, China, 21–23 June 2003.
34. Yuan, P.; Mao, Z.-Z.; Wang, F.-L. Endpoint prediction of EAF based on multiple support vector machines. *J. Iron Steel Res. Int.* **2007**, *14*, 20–24. [CrossRef]
35. Zhaoyi, L.; Zhi, X.; Hongji, M. Prediction model of end-point phosphorous in converter based on cluster analysis and gray theory. In Proceedings of the 2008 7th World Congress on Intelligent Control and Automation, Chongqing, China, 25–27 June 2008.
36. Wang, H.-B.; Xu, A.-J.; Ai, L.-X.; Tian, N.-Y. Prediction of endpoint phosphorus content of molten steel in BOF using weighted K-means and GMDH neural network. *J. Iron Steel Res. Int.* **2012**, *19*, 11–16. [CrossRef]
37. Wang, R.; Zhang, B.; Hu, C.; Liu, C.; Jiang, M. Modeling Study of Metallurgical Slag Foaming via Dimensional Analysis. *Metall. Mater. Trans. B* **2021**, *52*, 1805–1817. [CrossRef]
38. Qiu, D.; Fu, Y.-Y.; Zhang, N.; Zhao, C.-X. Research on relationship model of dephosphorization efficiency and slag basicity based on support vector machine. In Proceedings of the 2013 International Conference on Mechanical and Automation Engineering, Jiujang, China, 21–23 July 2013.
39. Liu, H.; Wang, B.; Xiong, X. Basic oxygen furnace steelmaking end-point prediction based on computer vision and general regression neural network. *Optik* **2014**, *125*, 5241–5248. [CrossRef]
40. Laha, D.; Ren, Y.; Suganthan, P.N. Modeling of steelmaking process with effective machine learning techniques. *Expert Syst. Appl.* **2015**, *42*, 4687–4696. [CrossRef]
41. He, F.; Zhang, L. Prediction model of end-point phosphorus content in BOF steelmaking process based on PCA and BP neural network. *J. Process Control* **2018**, *66*, 51–58. [CrossRef]
42. Elkoumy, M.M.; Fathy, A.M.; Megahed, G.M.; El-Mahallawi, I.; Ahmed, H.; El-Anwar, M. Empirical Model for Predicting Process Parameters during Electric Arc Furnace Refining Stage Based on Real Measurements. *Steel Res. Int.* **2019**, *90*, 1900208. [CrossRef]
43. Chang, S.; Zhao, C.; Li, Y.; Zhou, M.; Fu, C.; Qiao, H. Multi-channel graph convolutional network based end-point element composition prediction of converter steelmaking. *IFAC-PapersOnLine* **2021**, *54*, 152–157. [CrossRef]
44. Chen, C.; Wang, N.; Chen, M. Optimization of dephosphorization parameter in consteel electric arc furnace using rule set model. *Steel Res. Int.* **2021**, *92*, 2000719. [CrossRef]
45. Klimas, M.; Grabowski, D. Application of shallow neural networks in electric arc furnace modeling. *IEEE Trans. Ind. Appl.* **2022**, *58*, 6814–6823. [CrossRef]
46. Zhang, R.; Yang, J.; Wu, S.; Sun, H.; Yang, W. Comparison of the Prediction of BOF End-Point Phosphorus Content Among Machine Learning Models and Metallurgical Mechanism Model. *Steel Res. Int.* **2023**, *94*, 2200682. [CrossRef]
47. Zou, Y.; Yang, L.; Li, B.; Yan, Z.; Li, Z.; Wang, S.; Guo, Y. Prediction Model of End-Point Phosphorus Content in EAF Steelmaking Based on BP Neural Network with Periodical Data Optimization. *Metals* **2022**, *12*, 1519. [CrossRef]
48. Wang, R.; Mohanty, I.; Srivastava, A.; Roy, T.K.; Gupta, P.; Chattopadhyay, K. Hybrid method for endpoint prediction in a basic oxygen furnace. *Metals* **2022**, *12*, 801. [CrossRef]
49. Tomažič, S.; Andonovski, G.; Škrjanc, I.; Logar, V. Data-driven modelling and optimization of energy consumption in EAF. *Metals* **2022**, *12*, 816. [CrossRef]
50. Moosavi-Khoonsari, E.; Azzaz, R.; Hurel, V.; Jahazi, M.; Kahou, S.E. Controlling Minor Element Phosphorus in Green Electric Steelmaking Using Neural Networks. In Proceedings of the REWAS 2025 at TMS 2025 Annual Meeting & Exhibition (accepted), Las Vegas, NV, USA, 23–27 March 2025.
51. Reinicke, A.; Engbrecht, T.-N.; Schüttensack, L.; Echterhof, T. Application of an Artificial Neural Network for Efficient Computation of Chemical Activities within an EAF Process Model. *Metals* **2024**, *14*, 736. [CrossRef]
52. Zhou, K.-X.; Lin, W.-H.; Sun, J.-K.; Zhang, J.-S.; Zhang, D.-Z.; Feng, X.-M.; Liu, Q. Prediction model of end-point phosphorus content for BOF based on monotone-constrained BP neural network. *J. Iron Steel Res. Int.* **2022**, *29*, 751–760. [CrossRef]
53. Nenchev, B.; Panwisawas, C.; Yang, X.; Fu, J.; Dong, Z.; Tao, Q.; Gebelin, J.-C.; Dunsmore, A.; Dong, H.; Li, M.; et al. Metallurgical data science for steel industry: A case study on basic oxygen furnace. *Steel Res. Int.* **2022**, *93*, 2100813. [CrossRef]
54. Freuhan, J. *The Making, Shaping and Treating of Steel 11th Edition—Steelmaking and Refining Volume*; The AISE Steel Foundation: Pittsburgh, PA, USA, 1998.
55. Maia, T.A.; Onofri, V.C. Survey on the electric arc furnace and ladle furnace electric system. *Ironmak. Steelmak.* **2022**, *49*, 976–994. [CrossRef]
56. Singh, R. *Applied Welding Engineering: Processes, Codes, and Standards*; Butterworth-Heinemann: Oxford, UK, 2020; pp. 33–38.
57. Rathaba, L.P. *Model Fitting for Electric Arc Furnace Refining*; University of Pretoria (South Africa): Pretoria, South Africa, 2004.
58. Busa, N. *Optimization of Steelmaking Processes in an Electric ARC Furnace*; Purdue University: West Lafayette, IN, USA, 2023.
59. Kadkhodabeigi, M.; Tveit, H.; Johansen, S.T. Modelling the tapping process in submerged arc furnaces used in high silicon alloys production. *ISIJ Int.* **2011**, *51*, 193–202. [CrossRef]

60. Yang, X.-M.; Li, J.-Y.; Chai, G.-M.; Duan, D.-P.; Zhang, J. Critical evaluation of prediction models for phosphorus partition between CaO-based slags and iron-based melts during dephosphorization processes. *Metall. Mater. Trans. B* **2016**, *47*, 2302–2329. [CrossRef]
61. Nakamura, T.; Ueda, Y.; Yanagase, T. Optical Basicities in Some Oxide-Halide Systems. *ECS Proc. Vol.* **1987**, *1987*, 382. [CrossRef]
62. Nassaralla, C.; Fruehan, R. Phosphate capacity of CaO-Al_2O_3 slags containing CaF_2, BaO, Li_2O, or Na_2O. *Metall. Trans. B* **1992**, *23*, 117–123. [CrossRef]
63. Liu, Z.; Cheng, S.S.; Wang, L. Factors Influencing Dephosphorization of Low Carbon Steel in Converter. In *Materials Science Forum*; Trans Tech Publications: Stafa-Zurich, Switzerland, 2021.
64. Oh, M.K.; Park, J.H. Effect of fluorspar on the interfacial reaction between electric arc furnace slag and magnesia refractory: Competitive corrosion-protection mechanism of magnesiowüstite layer. *Ceram. Int.* **2021**, *47*, 20387–20398. [CrossRef]
65. Vieira, D.; Almeida, R.A.M.d.; Bielefeldt, W.V.; Vilela, A.C.F. Slag evaluation to reduce energy consumption and EAF electrical instability. *Mater. Res.* **2016**, *19*, 1127–1131. [CrossRef]
66. Li, F.; Li, X.; Zhang, Y.; Gao, M. Phosphate Capacities of CaO–FeO–SiO_2–Al_2O_3/Na_2O/TiO_2 Slags. *High Temp. Mater. Process.* **2019**, *38*, 50–59. [CrossRef]
67. Wang, Z.; Xie, F.; Wang, B.; Liu, Q.; Lu, X.; Hu, L.; Cai, F. The Control and Prediction of End-Point Phosphorus Content during BOF Steelmaking Process. *Steel Res. Int.* **2014**, *85*, 599–606. [CrossRef]
68. Probst, P.; Wright, M.N.; Boulesteix, A.L. Hyperparameters and tuning strategies for random forest. *Wiley Interdiscip. Rev. Data Min. Knowl. Discov.* **2019**, *9*, e1301. [CrossRef]
69. Cervantes, J.; Garcia-Lamont, F.; Rodríguez-Mazahua, L.; Lopez, A. A comprehensive survey on support vector machine classification: Applications, challenges and trends. *Neurocomputing* **2020**, *408*, 189–215. [CrossRef]
70. Roy, A.; Chakraborty, S. Support vector machine in structural reliability analysis: A review. *Reliab. Eng. Syst. Saf.* **2023**, *233*, 109126. [CrossRef]
71. Scikit Learn. RBF SVM Parameters. Available online: https://scikit-learn.org/stable/auto_examples/svm/plot_rbf_parameters.html (accessed on 1 September 2024).
72. IBM. What Is a Neural Network? Available online: https://www.ibm.com/topics/neural-networks#:~:text=Every%20neural%20network%20consists%20of,own%20associated%20weight%20and%20threshold (accessed on 1 September 2024).
73. Raiaan, M.A.K.; Sakib, S.; Fahad, N.M.; Al Mamun, A.; Rahman, M.A.; Shatabda, S.; Mukta, M.S.H. A systematic review of hyperparameter optimization techniques in Convolutional Neural Networks. *Decis. Anal. J.* **2024**, *11*, 100470. [CrossRef]
74. Bergstra, J.; Bengio, Y. Random search for hyper-parameter optimization. *J. Mach. Learn. Res.* **2012**, *13*, 281–305.
75. Karbowniczek, M.; Kawecka-Cebula, E.; Reichel, J. Investigations of the dephosphorization of liquid iron solution containing chromium and nickel. *Metall. Mater. Trans. B* **2012**, *43*, 554–561. [CrossRef]
76. Sigworth, G.K.; Elliott, J.F. The thermodynamics of liquid dilute iron alloys. *Met. Sci.* **1974**, *8*, 298–310. [CrossRef]
77. Yang, D.; Zhang, F.; Wang, J.; Yan, Z.; Pei, G.; Qiu, G.; Lv, X. Effect of Cr_2O_3 content on viscosity and phase structure of chromium-containing high-titanium blast furnace slag. *J. Mater. Res. Technol.* **2020**, *9*, 14673–14681. [CrossRef]
78. Ma, S.; Li, K.; Zhang, J.; Jiang, C.; Bi, Z.; Sun, M.; Wang, Z. Effect of MnO content on slag structure and properties under different basicity conditions: A molecular dynamics study. *J. Mol. Liq.* **2021**, *336*, 116304. [CrossRef]
79. Kawa, Y.; Mayani, H. Effect of alloying elements on the activity of phosphorous in molten iron. *Tetsu Hagane* **1982**, *68*.
80. Dovoedo, Y.; Chakraborti, S. Boxplot-based outlier detection for the location-scale family. *Commun. Stat. Simul. Comput.* **2015**, *44*, 1492–1513. [CrossRef]
81. Ratnasingam, S.; Muñoz-Lopez, J. Distance correlation-based feature selection in random forest. *Entropy* **2023**, *25*, 1250. [CrossRef] [PubMed]
82. Dahiru, T. P-value, a true test of statistical significance? A cautionary note. *Ann. Ib. Postgrad. Med.* **2008**, *6*, 21–26. [CrossRef] [PubMed]
83. Samarasinghe, S. *Neural Networks for Applied Sciences and Engineering: From Fundamentals to Complex Pattern Recognition*; Auerbach Publications: Boca Raton, FL, USA, 2016.

metals

MDPI

Article

Role of the Biogenic Carbon Physicochemical Properties in the Manufacturing and Industrial Transferability of Mill Scale-Based Self-Reducing Briquettes

Gianluca Dall'Osto, Davide Mombelli *, Sara Scolari and Carlo Mapelli

Dipartimento di Meccanica, Politecnico di Milano, Via La Masa 1, 20156 Milano, Italy; gianluca.dallosto@polimi.it (G.D.); sara.scolari@polimi.it (S.S.); carlo.mapelli@polimi.it (C.M.)
* Correspondence: davide.mombelli@polimi.it

Abstract: The recovery of iron contained in mill scale rather than iron ore can be considered a promising valorization pathway for this waste, especially if carried out through reduction using biogenic carbon sources. Nevertheless, the physicochemical properties of the latter may hinder the industrial transferability of such a pathway. In this work, the mechanical and metallurgical behavior of self-reduced briquettes composed of mill scale and four biogenic carbons (with increasing ratios of fixed carbon to volatile matter and ash) was studied. Each sample achieved mechanical performance above the benchmarks established for their application in metallurgical furnaces, although the presence of alkali compounds in the ash negatively affected the water resistance of the briquettes. In terms of metallurgical performance, although agglomeration successfully exploited the reduction by volatiles from 750 °C, full iron recovery and slag separation required an amount of fixed carbon higher than 6.93% and a heat treatment temperature of 1400 °C. Finally, the presence of Ca-, Al-, and Si- compounds in the ash was essential for the creation of a slag compatible with steelmaking processes and capable of retaining both phosphorus and sulfur, hence protecting the recovered iron.

Keywords: agglomeration; biogenic carbon; carbothermic reduction; iron recovery; mill scale; physicochemical properties; waste valorization

1. Introduction

Among the main industrial sectors, steelmaking is considered a pillar of societal development on a global scale, with its proliferation intimately linked to a growing and successful economy [1]. At present, steel is the most produced and consumed material for engineering applications, with a crude steel production of more than 1850 Mton in 2023 [2]. Nonetheless, the amount of CO_2 emissions from the steel production processes (5% and 7% of European and global human emissions, respectively) makes it a hard-to-abate sector with an urgent need to develop new low-carbon technologies and find new sources of materials that can make it independent from both mineral extraction from natural reserves and fossil carbon exploitation [3]. As a consequence, all major steel producers are currently following what could be called the transition from the primary steelmaking era, based on the use of sintering plant (SP), blast furnace (BF), and basic oxygen furnace (BOF) plants, to the electric steelmaking era, which consists of recycling scrap in electric arc furnaces (EAF) or electric melting furnaces (ESF), possibly coupled with direct reduction plants (DRP) [3]. Although this transition appears to be a possible solution to align with the EU climate goals, it will also result in increased demand for high-quality scrap and iron ore, the latter of which is needed for the production of direct reduced iron (DRI) or hot briquetted iron (HBI) to dilute the amount of harmful elements brought into the bath if low-grade scrap (i.e., scrap subject to two or three remelts over time) are used [4].

In this regard, the utilization of steel mill wastes and by-products rather than ores to produce clean iron could be a viable and cost-effective strategy to avoid mining. In fact, at a closer look, the steel industry has a wide range of iron oxide-bearing residues that have not yet been fully exploited, leaving several small material loops waiting to be closed [5]. Among various wastes and residues, mill scale (MS) is one of the most significant examples, as it consists of more than 90 wt.% iron oxides and is closely linked to the steel production, being generated during the hot working of steel and iron products and accounting for 2 wt.% of the steel produced. Depending on grain size and oil content, MS is used as feed in SP [6], sold cheaply to other sectors (e.g., construction [7], refractory [8], pigment [9]) or, in the worst case, disposed of in landfills, resulting in the loss of a source of clean iron and the need for economic expenditures for steel production. However, according to Iluţiu-Varvara et al. [10], it is estimated that approximately more than 30% of MS is still not fully valorized worldwide.

The first attempts of MS valorization date back to the mid-1960s, when Khodakovskii et al. [11] and Manukyan et al. [12] investigated the feasibility of its exploitation as a starting material to produce pure iron powder and sponge. Although this line of research has continued and expanded to the present day with studies based on carbothermal [13–15], aluminothermic [16–18], and gas-based (e.g., CO and H_2) reduction processes [19–21], the findings were not easily transferable to an industrial scale. The main reasons were due to the use of fossil carbon sources, unsuitable process parameters (e.g., temperature or time), economically expensive reducing elements (e.g., Al), or non-acceptable chemical and granulometric properties of the reduced iron particle for the introduction to the furnaces.

According to the BREF Best Available Techniques (BAT) Reference Document for Iron and Steel production [6], although most MS finds its main route of use/valorization in SP (along with dusts and sludges), the relative percentage in feed commonly stops at about 5–6% and only in rare cases reaches 10–20%. The main limitations are due to the negative impact that some elements (e.g., Zn, Pb, and chlorides) would have on the subsequent reintroduction of the sintered product into the BF. It is worth noting that the use of oily MS as feedstock in SP has also been studied, but even in these studies, the main difficulties were found to be the volatilization of hydrocarbon impurities (0.3–10 wt.%), which clogged the filter bags resulting in a fire hazard, increased dust production, and an overall decrease in productivity [22,23].

This was due to the significant amounts of particles smaller than 200 μm in MS, which nevertheless make it very attractive and suitable for the creation of agglomerates (e.g., briquettes or pellets) rather than its direct injection into furnaces [10].

As a parallel decarbonization route, the research on the total or partial replacement of fossil carbon sources with biogenic sources (e.g., biochar, biocoke, or hydrochar, generally labeled from now on as BC) has been highly active in recent years, due to the carbon neutrality associated with such kinds of materials. Several trials have been performed both in the primary and electric steelmaking routes on the actual feasibility of their exploitation as a coal substitute [24–27], carburizing agent [27,28], and foaming agent [29,30]. However, the inherent heterogeneity of the biochar physicochemical properties and general chemistry, resulting from both the starting biomass (e.g., woody matrix, food waste, sludge) and the production process (e.g., pyrolysis, torrefaction, hydrothermal carbonization), provided erratic results. On the one hand, satisfactory results were achieved when the properties were comparable to those of coke and coal; on the other hand, the early and strong gasification of low fixed carbon biochars inhibited their effectiveness and in some cases led to overheating and the subsequent shutdown of the dedusting system [31,32]. Nevertheless, the gasification of the low-grade BC could be beneficial because volatile evolution can contribute, to some extent, to the reduction of higher iron oxidation states (hematite and magnetite) at temperatures below 800 °C [33–35]. Despite the significant potential and environmental benefits of applying BC sources as reducing agents for iron recovery through the carbothermal reaction, there is currently a significant disproportion of studies exploiting iron ores as oxide-bearing materials [36–42] rather than MS [43–45].

Following the paradigms of waste utilization for mineral resource conservation and fossil carbon independence, this work focuses on the mechanical and metallurgical characterization of MS/BC self-reducing briquettes, with specific attention being paid to the effect of the physicochemical properties of the biogenic carbon source on the reduction behavior. The main challenges addressed are the achievement of the required characteristics for the introduction of such briquettes into the main steelmaking furnaces (e.g., BF, EAF, ESF, cupola furnaces, fluidized bed reactors, and rotary hearth furnaces), as well as the effective recovery of iron even when using low-grade BC. For this purpose, MS was mixed with four BC matrices, and, maintaining the low environmental impact of the process, an organic binder (corn starch) was exploited as a binding agent. Finally, although this study focuses on understanding the behavior of low- to high-grade MS/BS agglomerates, the results are also relevant to the establishment and the strengthening of the industrial symbiosis of the steel sector with major biomass-producing sectors (e.g., wooden/furniture, agricultural, and wastewater treatment plants).

2. Materials and Methods

2.1. Materials and Characterization

The briquettes were produced through agglomeration of the iron oxide-bearing matrix, rolling mill scale (MS), and four biogenic carbon matrices with increasing volatile matter content used as reducing agents and related to the wood, agriculture, and wastewater treatment sectors, namely, commercial wood pellets pyrolyzed at 750 °C for 25 min (WC), agricultural-derived olive pomace pyrolyzed at 750 °C and 350 °C for 25 min (OP750 and OP350), and sewage sludge hydrothermally carbonized at 210 °C at 20 bar (HC).

Prior to agglomeration, the MS was characterized by wavelength-dispersive X-ray fluorescence (WD-XRF) and X-ray diffraction (XRD) analysis to evaluate the chemical composition and mineralogical composition, the latter quantified by Rietveld method. The summary of the results is given in Table 1, and the XRD pattern is shown in the supplementary material (Figure S1).

Table 1. Chemical and mineralogical composition of mill scale (wt.%).

Wavelength-Dispersive X-ray Fluorescence									Rietveld Analysis		
Al_2O_3	CaO	Cr_2O_3	CuO	Fe_2O_3	MgO	MnO	NiO	SiO_2	Wustite	Magnetite	Hematite
0.30	0.32	0.45	0.10	96.95	0.07	1.13	0.08	0.60	60	30	10

The four reducing agents were characterized before agglomeration in terms of (i) percentage of fixed carbon, volatile, and ash by proximate analysis [46,47]; (ii) amount of total carbon and sulfur content by LECO analysis; (iii) activation energy by Kissinger–Akahira-Sunose (KAS) method [48]; (iv) water contact angle by sessile drop method [49]; and (v) mineralogical composition by XRD analysis. The summary of the physicochemical properties is given in Table 2; the XRD patterns of the reducing agents are given in the supplementary material (Figure S2).

Table 2. Proximate analysis, total carbon and sulfur content, activation energy, and water contact angle of reducing agents (WC: wood pellets pyrolyzed at 750 °C, OP750: olive pomace pyrolyzed at 750 °C, OP350 olive pomace pyrolyzed at 350 °C, HC: sewage sludge hydrothermally carbonized at 210 °C, C_{fix}: fixed carbon, VM: volatile matter, C_{tot}: total carbon, S: sulfur, E_a: activation energy, db: dry basis).

	Proximate Analysis (wt.%$_{db}$)			C_{tot} (wt.%)	S (wt.%)	Activation Energy (kJ mol^{-1})	Contact Angle (deg.)
	C_{fix}	VM	Ash				
WC	83.81	12.58	3.61	93.51	0.001	110.45	71.75
OP750	31.14	51.9	16.96	69.87	0.109	86.76	52.02
OP350	21.22	62.55	16.22	65.23	0.059	74.52	146.98
HC	0.73	69.95	29.32	42.67	0.589	61.61	157.84

2.2. Powder Briquetting

The materials were ground separately in a planetary ball mill with zirconium balls; subsequently, the mill scale was sieved to very fine sand (63–125 μm) and the reducing agents to coarse silt (less than 63 μm) to maximize the packing factor and reaction kinetics [50,51].

Four different recipes were obtained by adjusting the amount of reducing agent introduced to the stoichiometric carbon required for the reduction of all iron oxide contained in the mill scale (C_{tot}/FeO_{eq} equal to 0.2). The amount of MS in the recipe and that of fixed carbon, volatile matter, and ash (contributed by the reducing agent) are given in Table 3.

Table 3. Recipe composition (MS: mill scale, WC: wood pellets pyrolyzed at 750 °C, OP750: olive pomace pyrolyzed at 750 °C, OP350 olive pomace pyrolyzed at 350 °C, HC: sewage sludge hydrothermally carbonized at 210 °C, C_{fix}: fixed carbon, VM: volatile matter, db: dry basis).

	Mill Scale	C_{fix}	VM	Ash
	(wt.%$_{db}$)			
MS/WC	82.38	14.77	2.22	0.64
MS/OP750	77.75	6.93	11.55	3.77
MS/OP350	76.53	4.98	14.68	3.81
MS/HC	68.08	0.23	22.32	9.36

Corn starch was used as organic binder and dosed to correspond to 5 wt.% of the dry MS/BC pre-mix. To obtain a binder with a retrogradation rate of 30%, considered as the percentage able to provide the best mechanical performance to the agglomerates, distilled water (starch/water ratio of 1:6 by weight) was added to the starch and heated at 80 °C for 50 min [51]. The binder was immediately added to the recipe and then pressed with a constant speed of 20 mm min^{-1} until 40 MPa and maintained for 2 min, using a modified MTS Exceed Series 40 (MTS Systems, Eden Prairie, MN, USA) uniaxial tensile test machine (specific details of the machine are given in an earlier study by the authors [52]).

Cylindric briquettes of 20 mm in diameter, 20 mm in height, and about 15 g in mass were produced and dried for 14 days, during which the mass and volume of each briquette was monitored to evaluate the green and cured apparent density. To obtain a clear estimation of the volume, the samples were digitally processed into 3D spatial data by means of Agisoft Metashape photogrammetric processing software (version 1.8.5, Agisoft LLC, St. Petersburg, Russia).

2.3. Mechanical Characterization

The briquettes were mechanically characterized through impact resistance, cold compression, and water immersion resistance tests. Three replications were performed for each test to verify the reproducibility of the results.

Impact resistance was used to simulate falling briquettes inside a furnace; the ASTM D440-07 (2019) [53] was used as a guideline. Samples were dropped from a height of 1.63 m inside a tube with a vessel and stopped in case the briquettes withstood up to 10 falls or in case of premature failure. The impact resistance index (IRI) was calculated according to (1) whereas the detached material after each fall was collected, sieved, and weighed to estimate the size stability factor (*s*) according to (2).

$$\text{IRI} = \frac{n}{N} \times 100 \tag{1}$$

$$s(\%) = \sum_i wt_i \cdot \frac{\textit{Sieve Opening}_i}{\textit{Biggest Sieve Opening}} \times 100 \tag{2}$$

where n is the number of drops prior to failure, N is the number of pieces detached weighing 5% or more of the starting mass, and wt_i is the mass fraction of powder with respect to the starting mass retained in the *ith* sieve; specifically, the apertures used were 6.7 mm, 5.6 mm, 4 mm, 2 mm, 1 mm, 0.5 mm, and 0.125 mm.

Cold compression test was used to simulate the stresses at which the briquettes are subjected during the transportation or inside the furnace, the BS ISO 4700:2015 standard was used as a guideline [54]. The sample was pre-loaded between 2 flat plates at 30 N and pressed at a constant speed of 15 mm min^{-1} until the load falls by 50% of the ultimate compressive strength (UCS), or when the gap between the plates was reduced by more than 50% of the briquette diameter (10 mm).

The water resistance test was used to simulate the behavior of briquettes when exposed to rain or high humidity conditions that may adversely affect the quality of the densified product; the procedure described by Richards [55] was used as a guideline. The sample was fully immersed in distilled water at room temperature, removed after 600, 1200, and 1800 s, cleaned of surface moisture, and weighed each time. In addition, the visual appearance of the samples was constantly monitored by camera. The water resistance index (WRI) was calculated at 1800 s according to (3).

$$\mathrm{WRI}(\%) = \left(1 - \frac{w_{1800\mathrm{s}} - w_0}{w_0}\right) \times 100 \quad (3)$$

where w_{1800s} and w_0 are the mass of the sample after 1800 s and the initial one, respectively.

2.4. Metallurgical Performances Characterization

2.4.1. Reduction Behavior

The samples were thermally treated at four temperatures (750, 900, 1050 and 1200 °C) under inert atmosphere to evaluate the swelling, mineralogical evolution mass loss and degree of reduction subsequent to the heating.

The thermal cycles were carried in a Nabertherm LHT 02/17 LB lift-bottom furnace (Nabertherm, Lilienthal, Germany) in which a crucible containing the sample was inserted. The crucible was made in alumina and, to prevent any interaction with the briquette, a second graphite crucible was used as sample container. Since no variation of mass was observed in the graphite crucible in the experimental trials, it was concluded that it did not affect the results. To ensure an inert atmosphere, an argon flow (10·Nl h^{-1}) was directly blown inside the alumina crucible during the whole treatment duration. Specifically, after a fist maintenance at room temperature of 30 min, required to saturate the crucible volume with argon, the target temperature was reached using a heating rate of 100 °C min^{-1} and maintained for 15 min; finally, the crucible was naturally cooled down in the oven and extracted below 300 °C. The crucible configuration is given in Figure 1.

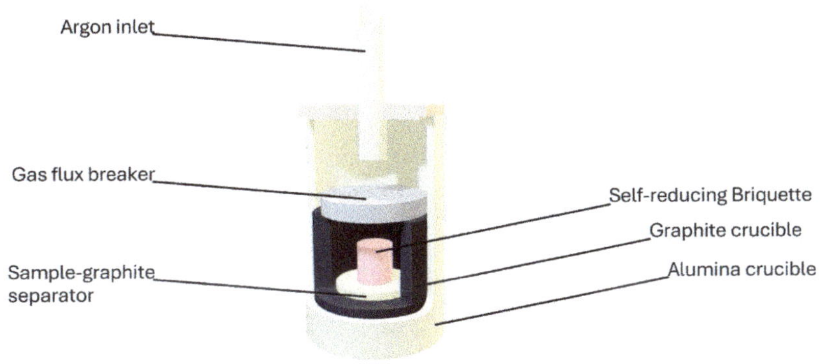

Figure 1. Schematics of the crucible used for the thermal treatment testing.

The swelling index was evaluated considering the apparent volume of the briquettes assuming the absence of internal porosities and according to (4).

$$\text{Swelling}\ (\%) = \frac{V_i - V_0}{V_0} \times 100 \quad (4)$$

where V_i and V_0 are the sample volume after the thermal cycle and the initial one, respectively. To obtain a clear estimation of the volume, the samples were digitally processed into 3D spatial data by means of Agisoft Metashape photogrammetric processing software (version 1.8.5, Agisoft LLC, St. Petersburg, Russia).

To focus on the role of the physicochemical properties of the reducing agents alone, the mass loss of the briquettes was evaluated by excluding the mass of the binder, assuming that it does not contribute largely to the reduction of iron oxides and calculated according to (5).

$$\Delta\text{m}\ (\%) = \frac{w_i - w_0}{w_0} \times 100 \quad (5)$$

where w_i and w_0 are the sample mass after the thermal cycle and the initial one, respectively (both masses were evaluated considering only the iron-bearing matrix and the reducing agent).

As pointed out by Vitikka et al. [43], carbon gasification during heating is responsible for the degree of reduction (DoR) bias in the case of self-reducing agglomerates with biogenic carbon sources. Consequently, to overcome this shortcoming, a modified version of BS ISO 11258:2015 [56], reported in (6), was used in this study to also take it into account.

$$\text{DoR}\ (\%) = \frac{w_i - w_0}{w_{th} - w_0} \times 100 \quad (6)$$

where w_{th} is the theoretical mass loss, calculated as the sum of the total expected losses from CO evolution (related to the reduction of oxides) and that of the removal of non-carbon volatile matter (related to the gasification during heating). Specifically, the former was evaluated by considering the complete conversion of C_{tot} present in the sample into CO, while the latter from the proximate analysis results as the difference between volatile matter and that of C_{tot} and C_{fix} ($VM - (C_{tot} - C_{fix})$).

Mineralogical evolution was determined by XRD analysis using a Rigaku Smartlab SE diffractometer (Rigaku Corporation, Tokyo, Japan) with CuKα radiation (λ = 1.54 Å) equipped with XRF suppression 1D detector (D/Tex 250). The samples were scanned from 20 to 120° 2θ at 1° min^{-1} with a step size of 0.02 ° and rotated at 120 rpm to decrease texture contribution.

2.4.2. Smelting Behavior and Iron Recovery

A fifth thermal treatment at 1400 °C, still under inert atmosphere, was conducted to investigate the feasibility of achieving the smelting of the reduced iron and its microstructure. In cases in which the briquette maintained its agglomerate morphology, the samples were cut in half to expose the cross section, mounted using a phenolic resin, and polished up to 1 µm. Otherwise, the reduced iron droplets were directly mounted and polished.

Metallographic analysis was carried out through optical microscopy whereas a Zeiss Sigma 300 Field Emission Gun Electronic Scanning Microscope (FEG-SEM) equipped with an Oxford Xmax Ultim 65 Energy-Dispersive X-ray Spectroscopy (EDS) probe (Carl Zeiss AG, Jena, Germany) was exploited to map the phase distribution at the end of the reduction process.

Finally, the degree of metallization (DoM) at 1400 °C was determined by quantitative metallurgy [57] by the post-processing of five backscatter electron micrographs taken stochastically along the polished surface of the sample and calculated according to (7).

$$\text{DoM}(\%) = \frac{Fe_M}{Fe_T} \times 100 = \frac{n_{Fe} \cdot \rho_{Fe}}{n_{Fe} \cdot \rho_{Fe} + n_{FeO} \cdot \rho_{FeO} \cdot \chi} \times 100 \quad (7)$$

where Fe_M is the metallic iron, Fe_T is the total iron, n_{Fe} is the number of pixels attributed to the metallic iron, ρ_{Fe} is the metallic iron density, n_{FeO} is the number of pixels attributed to the oxide iron, ρ_{FeO} is the oxide iron density, and χ is the weight percentage of iron in the specific oxide observed.

3. Results

3.1. Briquette Mechanical Performances

The as-produced briquettes (green) were let dry at room temperature to allow their curing and density stabilization. The visual appearance of the briquettes after 14 days (cured) is shown in Figure 2.

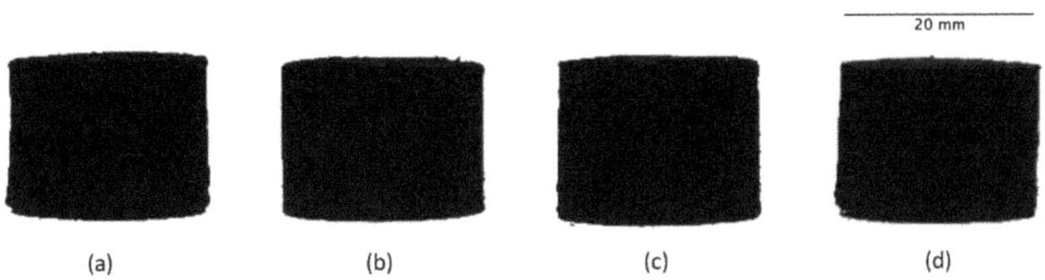

Figure 2. Appearance of the briquettes after 14 days of curing (**a**) MS/WC, (**b**) MS/OP750, (**c**) MS/OP350, and (**d**) MS/HC.

Although each briquette was able to self-sustain after the curing period, the surface of MS/WC and MS/OP750 samples differed strongly from that of MS/OP350 and MS/HC. On the one hand, the first two recipes were able to produce an agglomerate with a compact and smooth surface, while on the other hand, the MS/OP350 and MS/HC briquettes were characterized by the presence of several longitudinal cracks. Their presence is synonymous of a poor powder compaction, which makes the agglomerate unable to accommodate the shear stresses generated during the demolding step [58]. The differences in compaction behavior may also be associated with the presence of graphitic and amorphous carbon in WC and OP750 (Figure S2a,b), which acted as a lubricant during both the compression and demolding stages, reducing friction against the mold walls and the related shear stresses, in contrast to cellulose or hemicellulose in OP350 and HC (Figure S2c,d). Furthermore, it is worth noting that during the curing period, the hydrophobic nature of OP350 and HC (Table 2) may have favored the removal of moisture from the binder resulting in a widening of the cracks themselves. The apparent density of the briquettes and their mechanical properties are given in Table 4.

Table 4. Apparent density and mechanical properties of the briquettes (standard deviation in parenthesis).

	MS/WC	MS/OP750	MS/OP350	MS/HC
Apparent density, green (g cm^{-3})	2.48 (0.06)	2.69 (0.04)	2.33 (0.06)	2.35 (0.06)
Apparent density, cured (g cm^{-3})	2.34 (0.04)	2.57 (0.05)	2.16 (0.05)	2.24 (0.07)
Number of drop/1.63 m	10.0 (0.0)	9.5 (0.5)	3.0 (2.0)	7.7 (1.2)
Impact resistance index	1000.0 (0.0)	766.7 (230.2)	75.0 (22.5)	230.0 (4.8)
Size stability (%)	99.60 (3.64)	96.02 (1.63)	76.01 (10.52)	86.33 (6.11)
Ultimate compressive strength (MPa)	15.39 (0.19)	12.56 (0.74)	3.91 (0.56)	6.33 (0.74)
Water resistance index (%)	90.41	Failed before 600 s	Failed before 600 s	80.45

Each agglomerate underwent a decrease in apparent density from a green to cured state due to starch retrogradation, reaching a final value between 2.16 and 2.57 g cm^{-3} [51]. Although these values do not pose significant problems in their handling, transport, and storage, the slightly lower apparent densities, compared with that of EAF slag (2.98 g cm^{-3} at 1873 K), would result in their floating on the melt, with a consequent loss in iron recovery capacity if introduced directly into the furnace [59]. One possible solution is their placement in the lower layers of the charge to allow reduction during the first heating phase and the parallel increase in density due to the conversion of iron oxide to metallic iron. Conversely, since in the BF, cupola, and rotary heart processes the charge is solid, this problem would be less impactful as well as in line with the quality reference values of iron oxide pellets used in the aforementioned furnaces (2–2.5 g cm^{-3}) [60].

The difference in the specific apparent density, both in the green and cured states, is mainly related to three recipe-specific properties: (1) the relative mass ratio between MS and the reducing agent—the greater the value the greater the density, (2) the volumetric particle size distribution—the smaller the overall particle size the lower the density, and (3) the amount of inorganic content (ash) in the reducing agent—the greater the ash the higher the density [61]. The consequences of the specific recipe characteristics on the apparent density are most appreciable in the MS/WC and MS/OP750 briquettes comparison due to their compactness. Indeed, although the former was characterized by a greater amount of MS and a consequent particle size more oriented toward 125 μm, the greater amount of inorganic matter present in the latter was responsible for the higher density observed [62]. In addition, the lower hydrophilicity of WC enhanced the removal of moisture from the briquette during the curing period, leading to a greater decrease in density in the MS/WC briquette than in MS/OP750 (5.64% vs. 4.46%) [51]. The presence of cracks in the MS/OP350 and MS/HC briquettes distorted the apparent density measurements in the cured state, decreasing its value. Nevertheless, considering only the green state of the MS/OP350 and MS/HC briquettes, they were characterized by the lowest apparent densities among the samples, confirming the role of the relative mass ratio between the recipe powders and the volumetric particle size distribution discussed by Zhang et al. [61]. Finally, it is interesting to note that the direct comparison of the MS/OP350 and MS/HC cured briquettes showed a higher apparent density of the latter due to the higher amount of ash in the starting recipe.

The number of drops from a standstill to failure was, in ascending order of performance, 1 to 5 times/1.63 m for MS/OP350, 7 to 9 times/1.63 m for MS/HC, 9 to 10 times/1.63 m for MS/OP750, and 10 times/1.63 m for MS/WC. In addition, because this latter managed to remain intact until the tenth fall, its IRI resulted equal to 1000, while, despite the similar impact resistance of the MS/OP750 briquette, its IRI was found to be 766.7, due to its breakage during the last falls. The excellent impact resistance of the MS/WC and MS/OP750 samples is also reflected in their respective size stability of 99.60 and 96.02%, which means the low dispersion of fines after impact. Although the MS/HC and MS/OP350 briquettes showed the lowest IRI, namely 230 and 75, the values were still above the benchmark for handling and transporting briquettes for industrial applications (IRI > 50), with acceptable dispersion of fines after their failure (s > 75%) due to the binding effect given by corn starch, and with a number of drops before failure equal to or greater than that required by the iron oxide agglomerates used in BFs or shaft furnaces (drops before failure > 4) [60,63].

The difference in impact resistance between the MS/WC and MS/OP750 briquettes and the MS/OP350 and MS/HC briquettes is indirectly attributable to the physicochemical properties of the reducing agents. Indeed, due to the lower compactability of OP350 and HC with MS, the cracks and fissures on the final surface of the briquettes, as well as the possible presence of porosity within the matrix of the briquettes themselves, acted as sources of stress and points of failure during impact [61,64].

The compression results followed the identical order of impact strength performance: the highest UCS was observed for the MS/WC briquette (15.39 MPa), followed closely by the MS/OP750 one (12.56 MPa), while the MS/HC briquette (6.33 MPa) performed intermediately between the first two samples and the MS/OP350 briquette (3.91 MPa). Regardless, each briquette exceeded the reference value required for industrial storage and loading on conveyor belts (0.375 MPa) as well as that required for iron oxide agglomerates used in BFs or shaft furnaces (UCS > 7.9–9.5 MPa) [60,63].

The higher compressive strength of the MS/WC and MS/OP750 briquettes is consistent with the observations of El-Hussiny and Shalabi [65], who observed an increase in the UCS of briquettes proportional to the amount of MS present, at a constant amount of binder and agglomeration pressure.

Only the MS/WC and MS/HC briquettes were able to withstand immersion in water up to 1800 s, with a water absorption of 9.59 and 19.55%, respectively. According to Davies and Davies [66], the water resistance is inversely proportional to the amount of porosity of the briquette, which explained the lower water absorption of MS/WC briquette compared with the MS/HC briquette, despite the hydrophilic nature of WC compared with the hydrophobic of HC. On the one hand, the smooth surface of the MS/WC briquette prevented the infiltration of water, limiting the mass gain to the hydration of the superficial WC particles. On the other, the presence of cracks in the MS/HC briquette allowed the infiltration of water into the porosity, as suggested by the expulsion of bubbles from the briquette during the immersion (Figure S3a,d).

In contrast, both briquettes produced with OP as the reducing agent disintegrated rapidly upon contact with water (before 600 s of immersion), suggesting the presence of collateral properties of the reducing agents that affect water resistance. The main difference between the OP750 and OP350 reducing agents compared with WC and HC lies in the presence of water-soluble sylvite (KCl) (Figure S2b,c), which act as a catalyst for the dissolution of calcite ($CaCO_3$) when in contact with water [67]. Specifically, the dissolution of calcite increased the number of porosities present, exposing more of the hydrophilic OP750 particles to hydration, hence causing the briquette to swell severely to the point of failure (Figure S3b). Conversely, the hydrophobic nature of OP350 inhibited the particles' hydration and related swelling, confining the failure mechanisms to the loss of mechanical continuity due to the dissolution of sylvite and calcite (Figure S3c).

3.2. Metallurgical Performance Characterization

Figure 3 shows the mineralogical evolution of the briquettes after heating at 750, 900, 1050, and 1200 °C under Ar atmosphere.

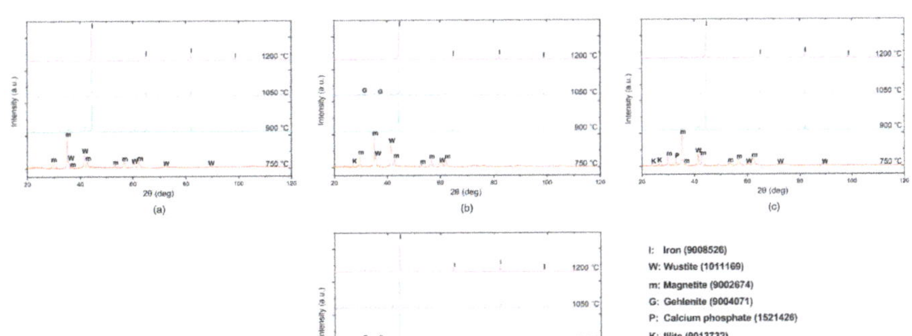

Figure 3. XRD pattern of the (**a**) MS/WC, (**b**) MS/OP750, (**c**) MS/OP350, and (**d**) MS/HC briquette after heating at 750, 900, 1050, and 1200 °C under Ar atmosphere (DB card number in parenthesis).

The XRD patterns showed a similar trend of the main reduction process (iron oxides to metallic iron) regardless of the reducing agent used. The MS/WC sample reduced at 750 °C was characterized by the presence of magnetite (Fe_3O_4) and wustite (FeO) only. Further to the latter, traces of illite ($KAl_2Si_4O_{10}(OH)_2$), calcium phosphate ($Ca_3(PO_4)_2$), hydroxyapatite ($Ca_5(PO_4)_3OH$), and gypsum ($CaSO_4 \cdot 2H_2O$) were observed in the MS/OP750, MS/OP350, and MS/HC briquettes and attributed to the decomposition/reaction between the compounds contained in the ash of the reducing agent and the minor elements of

MS. It is, however, of high interest to highlight that the MS/HC sample was the only one characterized by the presence of the iron peak, whose presence was attributed to the iron oxide reduction by the high amount of volatile contained in the recipe [35]. Though the presence of residual iron oxides, mainly in the form of wustite, was still observed at 900 °C, already at this temperature, the main diffraction peaks were those attributed to the metallic iron, highlighting the start of the reduction process. Furthermore, the presence of gehlenite (Ca_2Al_2SiO) in the MS/HC spectrum was due to the illite decomposition, at temperatures above 800 °C, and its catalyzing effect towards the calcining of gypsum [68,69]. Finally, at 1050 °C and 1200 °C, all the reduction patterns were characterized by the presence of metallic iron as the main iron phase, with traces of unreduced iron oxides associated either with post-reduction oxidation of the samples or with an incomplete reduction of the samples, mainly evident in the MS/WC and MS/HC samples.

The samples' swelling, mass loss, and DoR after heating at 750, 900, 1050, and 1200 °C under Ar atmosphere are shown in Figure 4. To better highlight the role of the physicochemical properties of the reducing agents on the metallurgical performance, the percentage of C_{tot}, volatile matter, and ash present in each recipe as well as the specific activation energy were reported in the graphs, too.

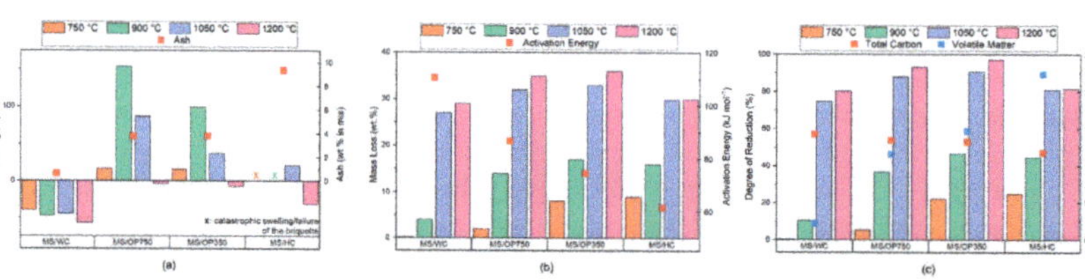

Figure 4. Comparison of the briquettes (**a**) swelling, (**b**) mass loss, and (**c**) degree of reduction after heating at 750, 900, 1050, and 1200 °C and effect of the specific reducing agent physicochemical properties.

The swelling behavior of the four briquette recipes (Figure 4a) agreed with the phenomenological description advanced by Reddy et al. [70]. Indeed, although the swelling is typically associated with (1) the transition from the highest to the lowest oxidation state of iron (Fe_2O_3 to FeO), (2) the reduction from wustite to metallic iron, and (3) the growth of such reduced iron fibers (whiskers), the evolution of slag and gases plays a key role in the swelling of the self-reducing aggregates [71]. Specifically, the slag evolution is intrinsically linked to the amount and chemical composition of the reducing agent ash, with alkali metal oxides (e.g., K_2O and Na_2O) considered to be the most detrimental to swelling behavior even when present in low percentages [51,70].

The MS/WC briquette was the only one characterized by negative swelling (shrinkage) regardless of heat treatment temperature, due to the almost null amount of ash. Furthermore, both the overall shrinkage trends and values corresponded to those observed by Reddy et al. [70] during the reduction of iron ore composites and spent mushroom substrate, with an increase in shrinkage between 750 and 900 °C (−40.02 and −47.55%), followed by a relative minimum at 1050 °C (−44.70%), and a final shrinkage once heated to 1200 °C (−56.92%). In contrast, in both MS/OP750 and MS/OP350, the potassium oxide liberation from the illite decomposition, as well as CO evolution due to MS reduction, increased the swelling value of the agglomerates from normal at 750 °C to abnormal at 900 °C. The higher value of the MS/OP750 briquette (153.62%) at 950 °C compared with that of MS/OP350 (99.02%) was associated with a higher concentration of potassium in the starting ash as well as the growth of a higher number of iron whiskers inside the agglomerate [43,71]. The increase in temperature allowed the sintering of reduced iron fibers in the MS/OP750 and MS/OP350 samples and the subsequent decrease in swelling at 1050 °C to shrinkage at 1200 °C (from 86.72 to −4.12% and 37.04 to −7.37%, respectively). Finally, the MS/HC briquette suffered most from the presence of alkali oxide and gas evolution during heating, undergoing catastrophic swelling at both 750 and 900 °C that caused the sample to fail at the end or even during heating. The small swelling of the MS/HC briquette at 1050 °C (20.73%) was associated with the lower amount of nucleated iron whiskers, as evidenced by the presence of wustite in the corresponding diffraction pattern, and hence the lower stresses against the briquette walls during their growth (Figure 3d).

Both the mass loss (Figure 4b) and the DoR (Figure 4c) followed a similar trend, allowing to interchange them for the description of the reduction behavior of the sample in the 750–1200 °C range. Specifically, the higher the temperature the higher the respective value; furthermore, considering the nature of the briquette recipes studied, the transformation of iron oxide to metallic iron is the main contributor to the mass loss (Figure 3). Accordingly, the significant mass loss observed at 750 °C only for the MS/OP350 and MS/HC briquettes (8.40 and 9.15%) was attributed to the gasification reactions occurring during heating [34,43]. This was especially evident when comparing the former with the MS/OP750 sample, which despite having an amount of volatile matter and C_{tot} in the recipe close to that of MS/OP350, the higher activation energy of the OP750 inhibited the initiation of gasification resulting in a lower mass loss (1.96%). In contrast, the MS/WC sample was considered a control case because of its almost null value of volatiles, which resulted in almost no mass loss (0.13%). Although, the shift from the reduction control by gas/volatile matter to carbon (occurring between 750 and 900 °C) was responsible for an appreciable mass loss of the MS/WC and MS/OP750 briquettes (1.96 and 13.65%), the former was still characterized by the lowest value due to the higher activation energy that inhibited the complete conversion of iron oxides, as confirmed by the presence of the wustite peaks in the corresponding diffraction pattern at 900 °C (Figure 3a) [72,73]. On the other hand, in the 900–1050 °C range, the WC was properly activated and able to carry out the oxide reduction with a sixfold increase in MS/WC mass loss, whereas the remaining briquettes increased their value twofold. Finally, a slight increase (<10%) in mass loss was observed at 1200 °C, regardless of the reducing agent used, with a final value in the 29–36% range, confirming the small appreciable difference between 1050 and 1200 °C from a mineralogical point of view.

To better highlight the sigmoidal trend of the DoRs versus the temperature (Figure 4c), the values were plotted together using a modified Avrami solution of the sigmoidal function [74], in which the time parameter was replaced by the reduction temperature (expressed in degrees Kelvin), and working under the assumption that the specific DoR reached the maximum value attainable in each thermal treatment, the linearized form plot is shown in Figure 5a. Furthermore, although the use of kinetic models to describe the behavior of the self-reducing agglomerate has been explored in the literature, to the authors knowledge, there are currently no studies on the correlation of the physicochemical properties of the reducing agent with respect to the model regression parameters [75,76]. Therefore, in this work, an attempt was made to correlate them to the amount of volatile matter and fixed carbon in each recipe. Specifically, the former was plotted against the constant parameter (k) and the latter against the slope of the regression curve (n); the plots are shown in Figures 5b and 5c, respectively.

Each regression curve showed a linear correlation coefficient (R-square) greater than 0.90, confirming the sigmoidal trend of DoR versus temperature. Similarly, the R-squares of volatile matter and fixed carbon with respect to $ln(k)$ and n were close to unity (0.96 and 0.98, respectively), further emphasizing their role on the proceeding in the reduction of mill scale agglomerates. On the one hand, the higher the volatile matter, the higher the regression constant and the DoR at low temperature due to gasification (below 1050 °C in this study); on the other hand, the higher the fixed carbon, the higher the slope of the regression curve and the ability to recover iron at high temperatures (above 1050 °C in this study).

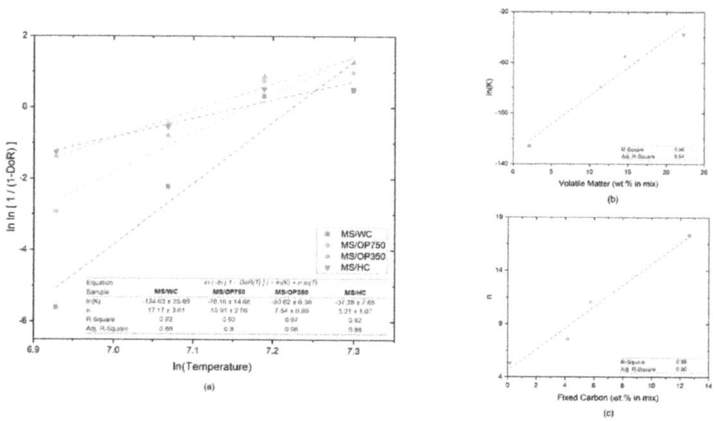

Figure 5. (**a**) Linearized form plots of the degree of reduction versus temperature, and relationship between (**b**) the regression constant and volatile matter in the recipe, and (**c**) the regression curve slope and fixed carbon in the recipe.

Smelting Behavior, Iron Recovery, and Industrial Transferability

Following the main purpose of extractive metallurgy, consisting of the recovery of valuable metals from metal-containing ores or sources and aiming at obtaining a secondary raw material in the form of clean iron to be reintroduced as additional feedstock into the main steelmaking furnaces, the four recipes thermally were treated at 1400 °C under Ar atmosphere for 15 min to evaluate their smelting behavior.

The visual appearance and metallography analysis of the samples after smelting is given in Figure 6. If the briquette was not able to achieve a full smelting, a metallographic preparation was carried out to expose the cross section, which was considered as most representative for evaluating the reduction morphology and porosity distribution.

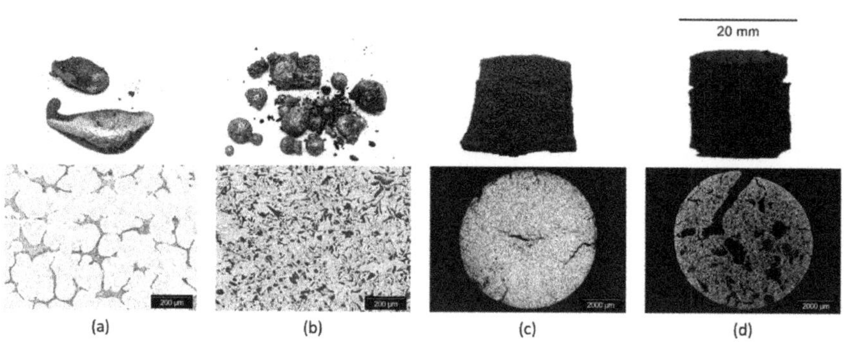

Figure 6. Visual appearance and metallographic analysis of the (**a**) MS/WC, (**b**) MS/OP750, (**c**) MS/OP350, and (**d**) MS/HC briquettes treated at 1400 °C.

On the one hand, the MS/WC and MS/OP750 samples were able to achieve a proper melting with a final microstructure typical of class III_{ISO} and V_{ISO} cast irons, respectively [77]. On the other hand, the MS/OP350 and MS/HC briquettes retained the agglomerated shape with a shrinkage of 65.84 and 57.25%, as well as a morphology similar to that observed by Sohn and Fruehan [78] following iron ore reduction by coal devolatilization. It is therefore interesting to further relate the slopes of the DoR regression curve (Figure 5c) to the smelting behavior of the briquettes. Based on the results, it can be stated that the heat treatment at 1400 °C of agglomerates with an n value (in the 750–1200 °C range) greater than 10.91 allows their smelting as well as the nucleation and growth of graphite flakes, with higher *n* values leading to coarser recovered iron droplets (Figure 6a,b), while the lower the value with respect to the threshold, the greater the amount of porosity and cavities left by gas evolution during the reduction process, and the less sintering and shrinkage (Figure 6c,d). Furthermore, the mass losses perfectly followed the amount of fixed carbon in the mixture, highlighting once again its major contribution at high thermal treatment temperatures. In ascending order, the values obtained were 36.58, 38.93, 40.24, and 46.27 percent from the MS/HC, MS/OP350, MS/OP750, and MS/WC samples, respectively.

SEM-EDS analysis was used to highlight the presence of harmful elements dispersed in the recovered iron matrix (e.g., P and S) and the chemical composition of the slag, with the latter only investigated for MS/OP350 and MS/HC briquettes; the micrographs are shown in Figure 7.

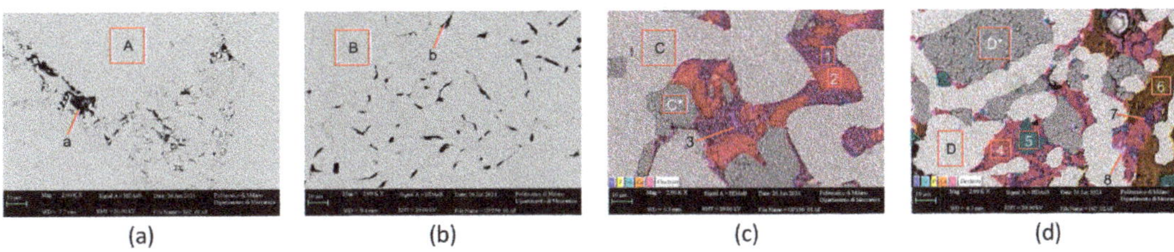

Figure 7. SEM micrograph of the (**a**) MS/WC and (**b**) MS/OP750 briquettes and SEM-EDS map of the cross section of the (**c**) MS/OP350 and (**d**) MS/HC briquettes.

SEM observation of the cross section associated with the EDS spectra (Table 5) confirmed the presence of graphite in the MS/WC and MS/OP750 iron matrices (Figure 7a,b), as well as the absence of sulfur and phosphorus in the latter, allowing the direct introduction of the reduced iron into metallurgical furnaces as secondary raw material for bath dilution without the need for substantial process modification or the risk of introducing harmful elements into the melt [79]. The SEM-EDS maps of the MS/OP350 and MS/HC briquettes (Figure 7c,d) showed a microstructure characterized by the presence of uniformly distributed islands of reduced and unreduced iron oxides throughout the surface, strongly linked to the slag phase. In particular, the MS/OP350 slag was characterized by a matrix of gehlenite together with calcium-olivine (Ca_2SiO_4) and secondarily precipitated crystals of iron-gehlenite ($Ca_2FeAlSiO_7$), within which phosphorus and sulfur were dissolved, resembling the mineralogy of high basicity BOF slags and that of self-reduced briquettes of flue dust and mill scale [80–82]. Further to gehlenite, a solid solution of olivine, kirschenite ($CaFeSiO_4$), iron-alumina spinel ($FeO{\cdot}Al_2O_3$), calcium phosphate ($Ca_3(PO_4)_2$), which is derived from the thermal decomposition of hydroxyapatite [83], and iron sulfide (FeS) constituted the mineralogy of the MS/HC sample slag.

From a practical point of view, the DoM achieved by the MS/WC, MS/OP750, and MS/OP350 briquettes (100, 100, and 82.32%, respectively), as well as the slag mineralogy of the latter, would allow their direct introduction into the main metallurgical furnaces as secondary raw material without affecting the specific process times and/or bath chemistries [45].

In contrast, the low DoM achieved by the MS/HC briquette (66.79%) would require a higher treatment temperature to achieve adequate iron recovery and be in line with the productivity constraints of metallurgical furnaces. For example, the DoM of self-reducing briquettes composed of iron ore fines, flue dust, and BOF/BF sludge increased from 69.70 to 97.95% when they were treated at 1500 instead of 1400 °C [84]. This temperature is achieved in BF, EAF, and cupola furnaces and would also be advantageous for the removal of sulfides observed in MS/HC briquette slag, as it would increase the desulfurization capacity of the slag itself, thus avoiding the need for a desulfurization agent after iron tapping/recasting (increased process costs) or a longer desulfurization period (increased process time) [85].

Table 5. EDS spectra of MS/WC, MS/OP750, MS/OP350, and MS/HC briquettes expressed in wt.% (A–D, C*–D* and a–b spectra are given as average values in the same phase).

Spectrum	C	O	Al	Si	P	S	Ca	Fe	Phase
A–D	3.57 [a]							96.43	Iron
C*–D*		24.24						75.76	Wustite
a–b	75.99	1.20						22.81	Graphite
1		38.83	14.05	11.54			27.55	8.03	Gehlenite
2		38.29		15.95			39.71	6.05	Calcio-olivine
3		37.39	2.82	15.56	0.2	0.71	24.41	18.91	Ferri-gehlenite
4		38.52		14.52	1.16		19.01	26.8	Kirschenite
5		34.17	22.87					42.96	Spinel
6		44.66			19.45		33.73	2.16	Calcium phosphate
7		44.66	18.12	10.33	4.18		11.49	11.23	Gehlenite
8		4.27		1.21		35	1.02	58.51	Iron sulfide

[a] Carbon derives from the iron carburation in the MS/WC and MS/OP750 samples.

Finally, considering the potential benefits that MS reduction through biogenic carbon would bring at the national level, the MS/OP750 recipe is the most promising. Indeed, the Italian olive production sector is the second largest in the world, with a crucial economic and social role in the southern regions, and is responsible for about 3 Mton of olive pomace per year as the main by-product; the latter is currently used as an energy source in incineration plants with unsatisfactory results, thus requiring the need for more promising valorization paths [86–88]. It is therefore possible to estimate that the creation of such a symbiosis of steel-agricultural waste valorization would not only have the potential to cover 18.70% of the 1.86 Mton of clean iron sources currently imported to Italy per year (updated data to 2022) [89], in the case of using the entire annual availability of mill scale, but also valorize 433 kton of raw pomace, with the latter being needed to achieve the 130 kton of OP750 (given a solid pyrolysis yield of 0.3 at 750 °C) to be agglomerated with MS as the reducing agent.

4. Conclusion

Aiming to deepen the knowledge of the role of physicochemical properties of biogenic carbon when agglomerated with mill scale, this work investigated the mechanical and metallurgical performance of four different recipes of self-reducing briquettes with increasing amounts of volatile matter and ash, as well as the parallel decrease of fixed carbon.

The main results of the present study can be summarized as follows:

- The mechanical performance are indirectly controlled by the hydrophilicity or hydrophobicity of the reducing agents used. Hydrophilic biogenic carbon allows the obtaining of agglomerates characterized by smooth surfaces and are highly packed, whereas hydrophobic biogenic carbon enhances the water expulsion from the agglomerate during the curing period leading to the formation of several cracks that inhibit the overall mechanical resistance, acting as low-resistance points. Still, the values achieved were comparable to or even better than that of pellets typically used in shaft furnaces (e.g., survival of 7 drops, CCS of 9.5 MPa, bulk density of 2 g cm^{-3}).
- The iron recovery is regulated by the amount of fixed carbon and volatile matter in the agglomerate. Specifically, though the briquetting is able to exploit the reduction of the iron oxides from the volatiles at 750 °C onward, even in the case of nearly null fixed carbon, an amount higher than 6.93% of fixed carbon and lower than 11.55% of volatile matter is required to fully recover the iron oxides of the agglomerate at 1400 °C, with a final microstructure associable to that of cast irons. Lower amounts of fixed carbon and higher volatile matter lead to the maintenance of the agglomerate morphology with the coexistence of reduced and unreduced iron as well as slag.
- The presence of alkali in the ash is highly detrimental for the agglomerates once immersed due to their dissolution and catalyzing effect towards the calcite hydration

which leads to briquette failure within 600 s. Furthermore, during heating, even a small amount of alkali can enhance the swelling in the 750–1050 °C range up to catastrophic values. Nevertheless, the presence of Ca-, Al-, and Si- compounds in the reducing agent ash allows the creation of a steelmaking close slag at 1400 °C, able to trap most of the phosphorus and sulfur, and protecting the recovered iron.

- Among the four biogenic carbon matrices, the use of wood chips or olive pomace pyrolyzed at 750 °C appears to be the most promising to recover the iron from mill scale and directly reintroduce it as secondary raw material in metallurgical furnaces when reduced at 1400 °C. On the contrary, slag separation processes and/or higher treatment temperatures are required for the recipes exploiting olive pomace pyrolyzed at 350 °C and hydrothermally carbonized sewage sludge to enhance the degree of metallization achieved (82.32% and 66.79%) up to industrial usable levels.

In conclusion, taking into account the future challenge for the valorization of mill scale that will arise due to the depletion of sintering plants and the shift to more electricity-focused iron and steel production, the agglomeration of such residue together with biogenic carbon sources could be a more than viable alternative solution for the recovery of the iron contained therein. Consequently, the next questions to be addressed will have to focus on understanding: (i) the optimal briquette charging conditions (e.g., quantity, time, relative position in the bucket, as unreduced or reduced) in the main furnaces (e.g., EAF, ESF, and BOF); (ii) any changes in the reduction behavior of virgin briquettes once interacting with the other charging materials, slag, and process period (reducing or oxidizing); and, most importantly, (iii) any alteration in the quality of the molten steel and the productivity of the furnaces themselves.

5. Patents

The experimental results shown and discussed in the present manuscript led to the setup of a process described and deposited in patent n. WO 2024/013653 A1 [90].

Supplementary Materials: The following supporting information can be downloaded at: https://www.mdpi.com/article/10.3390/met14080882/s1, Figure S1: XRD pattern of the mill scale; Figure S2: XRD patterns of the reducing agents (a) WC, (b) OP750, (c) OP350 and (d) HC; Figure S3: Temporal evolution of the (a) MS/WC, (b) MS/OP750, (c) MS/OP350 and (d) MS/HC briquettes during the water immersion resistance test.

Author Contributions: Conceptualization, G.D., D.M. and C.M.; methodology, G.D.; software, G.D. and S.S.; validation, G.D. and S.S.; formal analysis, G.D., D.M. and S.S.; investigation, G.D. and D.M.; resources, G.D. and S.S.; data curation, G.D. and D.M.; writing—original draft preparation, G.D.; writing—review and editing, G.D., D.M., S.S. and C.M.; visualization, G.D. and S.S.; supervision, D.M. and C.M. All authors have read and agreed to the published version of the manuscript.

Funding: (Italian version) Gli interventi oggetto della presente procedura sono finanziati a valere sulle risorse previste dal PNRR Missione 4 ("Istruzione e ricerca")—Componente 2 ("MICS-3A-ITALY-SPOKE 4")—Investimento 1.3 ("Partenariato Esteso Made in Italy Circolare e Sostenibile 3A-ITALY—Spoke 4 Smart and sustainable materials for circular and augmented industrial products and processes"), PE0000004, finanziato dall'Unione Europea-NextGenerationEU—CUP D43C22003120001. (English version) This study has been financed from the resources provided by the PNRR Mission 4 ("Education and Research")—Component 2 ("MICS-3A-ITALY-SPOKE 4")— Investment 1.3 ("Expanded Partnership Made in Italy Circular and Sustainable 3A-ITALY—Spoke 4 Smart and sustainable materials for circular and augmented industrial products and processes"), PE00004, funded by the European Union-NextGenerationEU—CUP D43C22003120001.

Data Availability Statement: The data presented in this study are available on request from the corresponding author (accurately indicate status).

Conflicts of Interest: The authors declare no conflicts of interest.

References

1. Liu, Y.; Li, H.; Huang, S.; An, H.; Santagata, R.; Ulgiati, S. Environmental and Economic-Related Impact Assessment of Iron and Steel Production. A Call for Shared Responsibility in Global Trade. *J. Clean. Prod.* **2020**, *269*, 122239. [CrossRef]
2. World Steel Association. *World Steel in Figures 2023*; World Steel Association: Bruxelles, Belgium, 2023.
3. International Energy Agency. *Iron and Steel Technology Roadmap*; International Energy Agency: Paris, France, 2020.
4. Dall'Osto, G.; Mombelli, D.; Mapelli, C. Consequences of the Direct Reduction and Electric Steelmaking Grid Creation on the Italian Steel Sector. *Metals* **2024**, *14*, 311. [CrossRef]
5. Colla, V.; Branca, T.A.; Pietruck, R.; Wölfelschneider, S.; Morillon, A.; Algermissen, D.; Rosendahl, S.; Granbom, H.; Martini, U.; Snaet, D. Future Research and Developments on Reuse and Recycling of Steelmaking By-Products. *Metals* **2023**, *13*, 676. [CrossRef]
6. BAT. *Best Available Techniques—Reference Document for Iron and Steel Production: Industrial Emissions Directive 2010/75/EU: Integrated Pollution Prevention and Control*; Joint Research Centre of the European Commission: Brussels, Belgium, 2012.
7. Iluțiu-Varvara, D.-A.; Aciu, C.; Tintelecan, M.; Sas-Boca, I.-M. Assessment of Recycling Potential of the Steel Mill Scale in the Composition of Mortars for Sustainable Manufacturing. *Procedia Manuf.* **2020**, *46*, 131–135. [CrossRef]
8. Spiliotis, X.; Ntampegliotis, K.; Kasiteropoulou, D.; Lamprakopoulos, S.; Lolos, K.; Karayannis, V.; Papapolymerou, G. Valorization of Mill Scale Waste by Its Incorporation in Fired Clay Bricks. *Key Eng. Mater.* **2014**, *608*, 8–13. [CrossRef]
9. Touzi, N.; Horchani-Naifer, K. A Study on the Preparation and Characterization of Pigment Quality from Mill Scale Steel Wastes. *Environ. Sci. Pollut. Res.* **2023**, *31*, 40538–40553. [CrossRef]
10. Iluțiu-Varvara, D.-A.; Aciu, C.; Maria Mârza, C.; Sas-Boca, I.-M.; Tintelecan, M. Assessment of Recycling Potential of the Oily Mill Scale in the Steelmaking Industry. *Procedia Manuf.* **2018**, *22*, 228–232. [CrossRef]
11. Khodakovskii, V.R.; Zhornyak, A.F. Estimate of the Supply of Mill Scale for the Manufacture of Iron Powders. *Sov. Powder Metall. Met. Ceram.* **1965**, *4*, 505–510. [CrossRef]
12. Manukyan, N.V. Carbide-Thermic Method of Preparation of Iron Powder. *Sov. Powder Metall. Met. Ceram.* **1967**, *6*, 260–263. [CrossRef]
13. Eissa, M.; Ahmed, A.; El-Fawkhry, M. Conversion of Mill Scale Waste into Valuable Products via Carbothermic Reduction. *J. Metall.* **2015**, *2015*, 1–9. [CrossRef]
14. Martín, M.I.; López, F.A.; Torralba, J.M. Production of Sponge Iron Powder by Reduction of Rolling Mill Scale. *Ironmak. Steelmak.* **2012**, *39*, 155–162. [CrossRef]
15. Cho, S.; Lee, J. Metal Recovery from Stainless Steel Mill Scale by Microwave Heating. *Met. Mater. Int.* **2008**, *14*, 193–196. [CrossRef]
16. Bugdayci, M.; Alkan, M.; Turan, A.; Yücel, O. Production of Iron Based Alloys from Mill Scale through Metallothermic Reduction. *High Temp. Mater. Process.* **2018**, *37*, 889–898. [CrossRef]
17. Kallio, M. Use of the Aluminothermic Reaction in the Treatment of Steel Industry By-Products. *J. Mater. Synth. Process.* **2000**, *8*, 87–92. [CrossRef]
18. Khanna, R.; Konyukhov, Y.; Li, K.; Jayasankar, K.; Maslennikov, N.; Zinoveev, D.; Kargin, J.; Burmistrov, I.; Leybo, D.; Kravchenko, M.; et al. Innovative Transformation and Valorisation of Red Mill Scale Waste into Ferroalloys: Carbothermic Reduction in the Presence of Alumina. *Sustainability* **2023**, *15*, 16810. [CrossRef]
19. Benchiheub, O.; Mechachti, S.; Serrai, S.; Khalifa, M.G. Elaboration of Iron Powder from Mill Scale. *J. Mater. Environ. Sci.* **2010**, *1*, 267–276.
20. Bagatini, M.C.; Zymla, V.; Osório, E.; Vilela, A.C.F. Characterization and Reduction Behavior of Mill Scale. *ISIJ Int.* **2011**, *51*, 1072–1079. [CrossRef]
21. Gaballah, N.M.; Zikry, A.F.; Khalifa, M.G.; Farag, A.B.; El-Hussiny, N.A.; Shalabi, M.E.H. Production of Iron from Mill Scale Industrial Waste via Hydrogen. *Open J. Inorg. Non-Met. Mater.* **2013**, *03*, 23–28. [CrossRef]
22. Domalski, E.S.; MacCrehan, W.A.; Moody, J.R.; Tewari, Y.B.; Walker, J.A. Characterization of Millscale Steel Wastes. National Bureau of Standards: Washington, DC, USA, 1983.
23. Umadevi, T.; Brahmacharyulu, A.; Karthik, P.; Mahapatra, P.C.; Prabhu, M.; Ranjan, M. Recycling of Steel Plant Mill Scale via Iron Ore Sintering Plant. *Ironmak. Steelmak.* **2012**, *39*, 222–227. [CrossRef]
24. Biochar for a Sustainable EAF Steel Production (GREENEAF2)—Publications Office of the EU. Available online: https://op.europa.eu/en/publication-detail/-/publication/7198c147-22b2-11e9-8d04-01aa75ed71a1/language-en (accessed on 6 July 2023).
25. Sustainable EAF Steel Production (GREENEAF)—Publications Office of the EU. Available online: https://op.europa.eu/en/publication-detail/-/publication/e7dc500c-82de-4c2d-8558-5e24a2d335fb/language-en (accessed on 6 July 2023).
26. Meier, T.; Hay, T.; Echterhof, T.; Pfeifer, H.; Rekersdrees, T.; Schlinge, L.; Elsabagh, S.; Schliephake, H. Process Modeling and Simulation of Biochar Usage in an Electric Arc Furnace as a Substitute for Fossil Coal. *Steel Res. Int.* **2017**, *88*, 1600458. [CrossRef]
27. He, X.-M.; Yi, S.; Fu, P.-R.; Zeng, X.-C.; Zhang, D.; Cheng, X.-H. Combustion Reactivity of Biochar and Char Generated from Co-Pyrolysis of Coal and Four Additives: Application in Blast Furnace. *J. Energy Eng.* **2017**, *143*, 04016023. [CrossRef]
28. Robinson, R.; Brabie, L.; Pettersson, M.; Amovic, M.; Ljunggren, R. An Empirical Comparative Study of Renewable Biochar and Fossil Carbon as Carburizer in Steelmaking. *ISIJ Int.* **2022**, *62*, 2522–2528. [CrossRef]
29. Cardarelli, A.; De Santis, M.; Cirilli, F.; Barbanera, M. Computational Fluid Dynamics Analysis of Biochar Combustion in a Simulated Ironmaking Electric Arc Furnace. *Fuel* **2022**, *328*, 125267. [CrossRef]

30. DiGiovanni, C.; Li, D.; Ng, K.W.; Huang, X. Ranking of Injection Biochar for Slag Foaming Applications in Steelmaking. *Metals* **2023**, *13*, 1003. [CrossRef]
31. Echterhof, T. Review on the Use of Alternative Carbon Sources in EAF Steelmaking. *Metals* **2021**, *11*, 222. [CrossRef]
32. Safarian, S. To What Extent Could Biochar Replace Coal and Coke in Steel Industries? *Fuel* **2023**, *339*, 127401. [CrossRef]
33. Konishi, H.; Ichikawa, K.; Usui, T. Effect of Residual Volatile Matter on Reduction of Iron Oxide in Semi-Charcoal Composite Pellets. *ISIJ Int.* **2010**, *50*, 386–389. [CrossRef]
34. Wei, R.; Cang, D.; Bai, Y.; Huang, D.; Liu, X. Reduction Characteristics and Kinetics of Iron Oxide by Carbon in Biomass. *Ironmak. Steelmak.* **2016**, *43*, 144–152. [CrossRef]
35. Bagatini, M.C.; Kan, T.; Evans, T.J.; Strezov, V. Iron Ore Reduction by Biomass Volatiles. *J. Sustain. Metall.* **2021**, *7*, 215–226. [CrossRef]
36. El-Tawil, A.; Ahmed, H.M.; El-Geassy, A.A.; Bjorkman, B. Effect of Volatile Matter on Reduction of Iron Oxide-Containing Carbon Composite. In Proceedings of the 54th Annual Conference of Metallurgists (COM 2015), Fairmont Royal York, Toronto, ON, Canada, 8 August 2015; pp. 1–14.
37. Das, D.; Anand, A.; Gautam, S.; Rajak, V.K. Assessment of Utilization Potential of Biomass Volatiles and Biochar as a Reducing Agent for Iron Ore Pellets. *Environ. Technol.* **2024**, *45*, 158–169. [CrossRef]
38. Sönmez, İ.; Şahbudak, K. Optimization of Sponge Iron (Direct Reduced Iron) Production with Box-Wilson Experimental Design by Using Iron Pellets and Lignite as Reductant. *Rev. Metal.* **2023**, *59*, e241. [CrossRef]
39. Murakami, T.; Takahashi, T.; Fuji, S.; Maruoka, D.; Kasai, E. Development of Manufacturing Principle of Porous Iron by Carbothermic Reduction of Composite of Hematite and Biomass Char. *Mater. Trans.* **2017**, *58*, 1742–1748. [CrossRef]
40. Liu, Z.; Bi, X.; Gao, Z.; Liu, W. Carbothermal Reduction of Iron Ore in Its Concentrate-Agricultural Waste Pellets. *Adv. Mater. Sci. Eng.* **2018**, *2018*, 1–6. [CrossRef]
41. Chuanchai, A.; Wu, K.-T. Potential of Pinewood Biochar as an Eco-Friendly Reducing Agent in Iron Ore Reduction. *ACS Omega* **2024**, *9*, 14279–14286. [CrossRef] [PubMed]
42. Ubando, A.T.; Chen, W.-H.; Ong, H.C. Iron Oxide Reduction by Graphite and Torrefied Biomass Analyzed by TG-FTIR for Mitigating CO_2 Emissions. *Energy* **2019**, *180*, 968–977. [CrossRef]
43. Vitikka, O.; Iljana, M.; Heikkilä, A.; Tkalenko, I.; Kovtun, O.; Koriuchev, N.; Shehovsov, D.; Fabritius, T. Effect of Biocarbon Addition on Metallurgical Properties of Mill Scale-Based Auger Pressing Briquettes. *ISIJ Int.* **2024**, *64*, 964–977. [CrossRef]
44. Khaerudini, D.S.; Chanif, I.; Insiyanda, D.R.; Destyorini, F.; Alva, S.; Pramono, A. Preparation and Characterization of Mill Scale Industrial Waste Reduced by Biomass-Based Carbon. *J. Sustain. Metall.* **2019**, *5*, 510–518. [CrossRef]
45. Bagatini, M.C.; Zymla, V.; Osório, E.; Vilela, A.C.F. Scale Recycling Through Self-Reducing Briquettes to Use in EAF. *ISIJ Int.* **2017**, *57*, 2081–2090. [CrossRef]
46. Dall'Osto, G.; Mombelli, D.; Pittalis, A.; Mapelli, C. Biochar and Other Carbonaceous Materials Used in Steelmaking: Possibilities and Synergies for Power Generation by Direct Carbon Fuel Cell. *Biomass Bioenergy* **2023**, *177*, 106930. [CrossRef]
47. *ASTM D1762-84*; American Society for Testing and Materials International Standard Test Method for Chemical Analysis of Wood Charcoal. ASTM: Conshohocken, PA, USA, 2009.
48. Lim, A.C.R.; Chin, B.L.F.; Jawad, Z.A.; Hii, K.L. Kinetic Analysis of Rice Husk Pyrolysis Using Kissinger-Akahira-Sunose (KAS) Method. *Procedia Eng.* **2016**, *148*, 1247–1251. [CrossRef]
49. Hussain, R.; Ghosh, K.K.; Garg, A.; Ravi, K. Effect of Biochar Produced from Mesquite on the Compaction Characteristics and Shear Strength of a Clayey Sand. *Geotech. Geol. Eng.* **2021**, *39*, 1117–1131. [CrossRef]
50. Mombelli, D.; Dall'Osto, G.; Trombetta, V.; Mapelli, C. Comparison of the Reduction Behavior through Blast Furnace Sludge of Two Industrial Jarosites. *J. Environ. Chem. Eng.* **2023**, *11*, 109360. [CrossRef]
51. Dall'Osto, G.; Mombelli, D.; Trombetta, V.; Mapelli, C. Effect of Particle Size and Starch Gelatinization on the Mechanical and Metallurgical Performance of Jarosite Plus Blast Furnace Sludge Self-Reducing Briquettes. *J. Sustain. Metall.* **2024**, *10*, 759–774. [CrossRef]
52. Mombelli, D.; Gonçalves, D.L.; Mapelli, C.; Barella, S.; Gruttadauria, A. Processing and Characterization of Self-Reducing Briquettes Made of Jarosite and Blast Furnace Sludges. *J. Sustain. Metall.* **2021**, *7*, 1603–1626. [CrossRef]
53. *ASTM D440-07*; American Society for Testing and Materials International Standard Test Method of Drop Shatter Test for Coal. ASTM: Conshohocken, PA, USA, 2019.
54. *BS ISO 4700:2015*; Iron Ore Pellets for Blast Furnace and Direct Reduction Feedstocks—Determination of the Crushing Strength. ISO: Geneva, Switzerland, 2015.
55. Richards, S.R. Physical Testing of Fuel Briquettes. *Fuel Process. Technol.* **1990**, *25*, 89–100. [CrossRef]
56. *BS ISO 11258:2015*; Iron Ores for Shaft Direct-Reduction Feedstocks. Determination of the Reducibility Index, Final Degree of Reduction and Degree of Metallization. ISO: Geneva, Switzerland, 2015.
57. Leino, T.; Taskinen, P.; Eric, R.H. Determination of Metallization Degree of Pre-Reduced Chromite with Image and Rietveld Analysis. *J. Min. Metall. Sect. B Metall.* **2020**, *56*, 289–297. [CrossRef]
58. Li, Z.; Zou, H. Optimization of Biomass Fuel Cold Briquetting Parameters Based on Response Surface Analysis. *J. Inst. Eng. (India) Ser. C* **2022**, *103*, 459–472. [CrossRef]
59. Seetharaman, S.; Teng, L.; Hayashi, M.; Wang, L. Understanding the Properties of Slags. *ISIJ Int.* **2013**, *53*, 1–8. [CrossRef]

60. Seetharaman, S. *Treatise on Process Metallurgy, Volume 3: Industrial Processes*; Elsevier: Amsterdam, The Netherlands, 2013; Volume 3.
61. Zhang, C.; Zhang, N.; Pan, D.; Qian, D.; An, Y.; Yuan, Y.; Xiang, Z.; Wang, Y. Experimental Study on Sensitivity of Porosity to Pressure and Particle Size in Loose Coal Media. *Energies* **2018**, *11*, 2274. [CrossRef]
62. Pang, L.; Yang, Y.; Wu, L.; Wang, F.; Meng, H. Effect of Particle Sizes on the Physical and Mechanical Properties of Briquettes. *Energies* **2019**, *12*, 3618. [CrossRef]
63. Richards, S.R. Briquetting Peat and Peat-Coal Mixtures. *Fuel. Process. Technol.* **1990**, *25*, 175–190. [CrossRef]
64. Suarez, J.A.; Luengo, C.A. Coffee Husk Briquettes: A New Renewable Energy Source. *Energy Sources* **2003**, *25*, 961–967. [CrossRef]
65. El-Hussiny, N.A.A.; Shalabi, M.E.H.E.H. A Self-Reduced Intermediate Product from Iron and Steel Plants Waste Materials Using a Briquetting Process. *Powder Technol.* **2011**, *205*, 217–223. [CrossRef]
66. Davies, R.M.; Davies, O.A. Effect of Briquetting Process Variables on Hygroscopic Property of Water Hyacinth Briquettes. *J. Renew. Energy* **2013**, *2013*, 429230. [CrossRef]
67. Finneran, D.W.; Morse, J.W. Calcite Dissolution Kinetics in Saline Waters. *Chem. Geol.* **2009**, *268*, 137–146. [CrossRef]
68. Araújo, J.H.d.; Silva, N.F.d.; Acchar, W.; Gomes, U.U. Thermal Decomposition of Illite. *Mater. Res.* **2004**, *7*, 359–361. [CrossRef]
69. Lu, D.; Chen, Q.; Li, C.; Gong, S. Effect of Potassium Feldspar on the Decomposition Rate of Phosphogypsum. *J. Chem. Technol. Biotechnol.* **2021**, *96*, 374–383. [CrossRef]
70. Reddy, D.S.; Chang, H.-H.; Tsai, M.-Y.; Chen, I.-G.; Wu, K.-T.; Liu, S.-H. Swelling and Softening Behavior of Iron Ore-Spent Mushroom Substrate Composite Pellets during Carbothermal Reduction. *J. Mater. Res. Technol.* **2023**, *22*, 1999–2007. [CrossRef]
71. Singh, M.; Björkman, B. Effect of Reduction Conditions on the Swelling Behaviour of Cement-Bonded Briquettes. *ISIJ Int.* **2004**, *44*, 294–303. [CrossRef]
72. Sarkar, S.B.; Ray, H.S.; Chatterjee, I. Kinetics of Reduction of Iron Ore—Coal Pellets. *J. Therm. Anal.* **1989**, *35*, 2461–2469. [CrossRef]
73. Bagatini, M.C.; Zymla, V.; Osório, E.; Vilela, A.C.F. Carbon Gasification in Self-Reducing Mixtures. *ISIJ Int.* **2014**, *54*, 2687–2696. [CrossRef]
74. Fanfoni, M.; Tomellini, M. The Johnson-Mehl- Avrami-Kohnogorov Model: A Brief Review. *Il Nuovo Cimento D* **1998**, *20*, 1171–1182. [CrossRef]
75. Yuan, X.; Luo, F.; Liu, S.; Zhang, M.; Zhou, D. Comparative Study on the Kinetics of the Isothermal Reduction of Iron Ore Composite Pellets Using Coke, Charcoal, and Biomass as Reducing Agents. *Metals* **2021**, *11*, 340. [CrossRef]
76. Chai, Y.; Fan, Y.; Li, Z.; Wu, J.; Zhang, Y.; Wang, Y.; Luo, G.; An, S. Kinetics of Reduction in Stages of Pellets Prepared from the Bayan Obo Iron Ore Concentrate. *ACS Omega* **2022**, *7*, 7759–7768. [CrossRef] [PubMed]
77. Friess, J.; Sonntag, U.; Steller, I.; Bührig-Polaczek, A. From Individual Graphite Assignment to an Improved Digital Image Analysis of Ductile Iron. *Int. J. Met.* **2020**, *14*, 1090–1104. [CrossRef]
78. Sohn, I.; Fruehan, R.J. The Reduction of Iron Oxides by Volatiles in a Rotary Hearth Furnace Process: Part III. The Simulation of Volatile Reduction in a Multi-Layer Rotary Hearth Furnace Process. *Metall. Mater. Trans. B* **2006**, *37*, 231–238. [CrossRef]
79. Ghosh, A.; Chatterjee, A. *Ironmaking and Steelmaking Theory and Practice*; PHI Learning Private Limited: New Delhi, India, 2008; Volume 20, ISBN 812033289X.
80. Rietmeijer, F.J.M.; Nuth, J.A.; Pun, A. The Formation of Mg,Fe-silicates by Reactions between Amorphous Magnesiosilica Smoke Particles and Metallic Iron Nanograins with Implications for Comet Silicate Origins. *Meteorit. Planet. Sci.* **2013**, *48*, 1823–1840. [CrossRef]
81. Liu, C.; Huang, S.; Blanpain, B.; Guo, M. Effect of Al2O3 Addition on Mineralogical Modification and Crystallization Kinetics of a High Basicity BOF Steel Slag. *Metall. Mater. Trans. B* **2019**, *50*, 271–281. [CrossRef]
82. Bagatini, M.C.; Fernandes, T.; Silva, R.; Galvão, D.F.; Flores, I.V. Mill Scale and Flue Dust Briquettes as Alternative Burden to Low Height Blast Furnaces. *J. Clean. Prod.* **2020**, *276*, 124332. [CrossRef]
83. Böhme, N.; Hauke, K.; Dohrn, M.; Neuroth, M.; Geisler, T. High-Temperature Phase Transformations of Hydroxylapatite and the Formation of Silicocarnotite in the Hydroxylapatite–Quartz–Lime System Studied in Situ and in Operando by Raman Spectroscopy. *J. Mater. Sci.* **2022**, *57*, 15239–15266. [CrossRef]
84. Mohanty, M.K.; Mishra, S.; Mishra, B.; Sarkar, S.; Samal, S.K. A Novel Technique for Making Cold Briquettes for Charging in Blast Furnace. *IOP Conf. Ser. Mater. Sci. Eng.* **2016**, *115*, 012020. [CrossRef]
85. Mombelli, D.; Mapelli, C.; Barella, S.; Gruttadauria, A.; Spada, E. Jarosite Wastes Reduction through Blast Furnace Sludges for Cast Iron Production. *J. Environ. Chem. Eng.* **2019**, *7*, 102996. [CrossRef]
86. Valenti, F.; Arcidiacono, C.; Chinnici, G.; Cascone, G.; Porto, S.M. Quantification of Olive Pomace Availability for Biogas Production by Using a GIS-based Model. *Biofuels Bioprod. Biorefining* **2017**, *11*, 784–797. [CrossRef]
87. CREA Consiglio per la Ricerca in Agricoltura e L'analisi Dell'economia Agraria. *Italian Agriculture in Figures 2022*; CREA Consiglio per la Ricerca in Agricoltura e L'analisi Dell'economia Agraria: Roma, Italy, 2022.
88. Vasileiadou, A.; Zoras, S.; Iordanidis, A. Bioenergy Production from Olive Oil Mill Solid Wastes and Their Blends with Lignite: Thermal Characterization, Kinetics, Thermodynamic Analysis, and Several Scenarios for Sustainable Practices. *Biomass Convers. Biorefinery* **2021**, *13*, 5325–5338. [CrossRef]

89. Federacciai. *La Siderurgia Italiana in Cifre: The Italian Steel Industry Key Statistics 2022*; Federacciai: Milano, Italy, 2023; Available online: https://federacciai.it/wp-content/uploads/2023/05/AssembleaAnnuale2023_Relazione-Annuale-2022.pdf (accessed on 28 July 2024).
90. Method for Direct Reduction of Iron Oxide-Based Material for The Production of Steel, Iron Sponge or Cast Iron. WIPO Patent Application WO/2024/013653, 18 January 2024.

metals

MDPI

Article

Application of an Artificial Neural Network for Efficient Computation of Chemical Activities within an EAF Process Model

Alexander Reinicke *, Til-Niklas Engbrecht, Lilly Schüttensack and Thomas Echterhof

Department of Industrial Furnaces and Heat Engineering, RWTH Aachen University, Kopernikusstraße 10, 52074 Aachen, Germany

* Correspondence: reinicke@iob.rwth-aachen.de

Abstract: The electric arc furnace (EAF) is considered the second most important process for the production of crude steel and is usually used for the melting of scrap. With the current emphasis on defossilization, its share in global steelmaking is likely to further increase. Due to the large production quantities, minor improvements to the EAF process can still accumulate into a significant reduction in overall energy and resource consumption. A major aspect in the efficient operation of the EAF is achieving beneficial slag properties, as the slag influences the composition of the steel and can reduce energy losses as well as the maintenance cost. In order to investigate the EAF operation, a dynamic process model is applied. Within the model, the chemical reactions of the metal–slag system are calculated based on the activities of the involved species. In this regard, multiple models for the calculation of the chemical activities have been implemented. However, depending on the chosen model, the computation of the slag activities can be computationally demanding. For this reason, the application of a neural network for the calculation of the chemical activities within the slag is investigated. The performance of the neural network is then compared to the results of the previously applied models by using the commercial software FactSage as a reference. The validation shows that the surrogate model achieves great accuracy while keeping the computation demand low.

Keywords: electric arc furnace; chemical activities; chemical equilibrium; regression; artificial neural network; FactSage; process model

1. Introduction

The electric arc furnace (EAF) is the second most important process for the production of crude steel [1]. Historically, it was primarily used for the melting of scrap metal. However, with the ongoing decarbonization of the steel industry, other input materials such as biogenic carbon carriers or direct reduced iron (DRI) are being used more frequently in the EAF [2,3]. Meanwhile, the energy costs as well as the requirements on the composition of the crude steel are increasing, posing an unprecedented challenge for the furnace operation. Due to the extreme condition within the arc furnace, the measurements, however, are highly complicated and cost-intensive. As a result, the available data are often limited to ingoing and outgoing energy and material flows, with no direct information on the occurring processes. In addition, there is often a significant delay between taking a sample of the steel and the determination of the result. A common way to analyze and improve the operation of metallurgical processes is, therefore, through the application of mathematical models. In doing so, the costs and risks connected with subsequent plant trials can be decreased. A comprehensive review of the available models of the EAF was published by Hay et al. [4] as well as Carlsson et al. [5]. Carlsson focused on statistical models, while Hay et al. discussed the purpose, modeling approach, and limitations of dynamic process models. A comprehensive and well-documented process model of the EAF was developed by Meier [6] based on the previous work by Logar et al. [7,8]. Within the model, the furnace is divided into several homogenous zones, including the liquid steel and slag, as well as the gas phase. The arc furnace process is simulated by calculating the energy and mass transfer between these zones. In this regard, the chemical reactions are a key aspect in simulating the EAF process, as they determine the composition of the molten steel, slag, and off-gas. Moreover, chemical reactions contribute a significant amount of energy to the process [9]. A well-conditioned slag also improves the stability of the arc, reduces the refractory corrosion, and reduces heat losses by building a slag coating on the furnace walls. In later stages of the process, it is advantageous to promote a foaming slag by injecting carbon and oxygen. The carbon reacts with the oxygen and oxides in the slag, forming CO, which leads to the foaming of the slag. The foaming slag encapsulates the arc and shields the furnace roof and walls from its thermal radiation, further reducing heat losses and preventing damage [10]. For the presented reasons, an accurate modelling of the slag phase as well as the reactions in the interphase between the melt and slag is detrimental for a simulation of the EAF process.

The reaction rates are subject to the melt and slag composition, and the activity of the species, as well as the reaction kinetics and the mass transfer within the adjacent zones. Although the slag activity is only calculated for a limited number of species, the computation takes up a significant fraction of the total runtime of the process model. The overall computation time is still short enough to allow for the real-time monitoring of the process. However, when running optimization tasks on multiple heats, the computation time can increase drastically. A possible future application of the model is the optimization of the operating chart in order to minimize energy and resource consumption as well as climate gas emissions while maintaining a desirable composition of the melt and a safe operation of the furnace. However, searching the high-dimensional solution space using the process model in its default state is not possible in a reasonable time. In this regard, a further reduction in the computation time is imperative. The main objective of this work is, therefore, to implement a surrogate model for the calculation of the activity of the reacting species by utilizing supervised machine learning. In this regard, neural networks allow for the representation of complex relationships between the input and output data. A shallow neural network is used, as it provides a sufficient approximation of the reference model, yet retains an efficient architecture. This way, by applying the surrogate model, the computation time of the EAF model can be significantly decreased. The described approach of substituting particularly demanding sub-models is not limited to the domain of the chemical activity calculation. It can also be applied to other model aspects or even different processes. The necessary training data can be generated by means of measurement or by the application of an accurate but computationally more demanding model.

2. Materials and Methods

In general, most relevant metallurgical reactions take place at the boundaries between immiscible phases. In the context of the EAF, those are primarily the injection zones of oxygen and carbon, as well as the interface between melt and slag [11]. Within these zones, the overall reaction rates are governed by the transport of reactants and products towards and from the interface, as well as the chemical reaction rates at the interface itself. However, with temperatures above the melting point of steel, a local equilibrium is often assumed for the interface, with the transport rates considered as the limiting factor [11]. Among others, this approach is used within the process models by MacRosty [12] and Hay [6,13]. The latter is the primary object of consideration for this work. Hay incorporates separate reactions zones for the sites of oxygen and carbon injection, as well as the metal–slag interface. However, for the purpose of demonstration, the surrogate model is only applied for the interaction zone between liquid metal and slag. Within the model, the reaction zone contains the entire liquid slag phase and a limited amount of melt. Consequently, in terms of mass transport, the reaction rates are only limited by the diffusion rates between the melt and the metal–slag interface ($\dot{m}_{\mathrm{Species\ i}}$) in accordance to Equation (1), where the current mass fractions of species i within the melt and interface are given by w_i^{melt} and $w_i^{\mathrm{interface}}$, respectively, and kd_i denoting empirical factors. The equilibrium composition within the interaction zone is governed by Equation (2). As such, the equilibrium constant $K_{eq,i}$ can be determined as a function of the standard Gibbs free energy change $-\Delta G^0$, temperature T, and the gas constant R, as well as the activity a_i and the stoichiometric coefficient v_i for each species. The equilibrium concentration of the elements in the interface can then be calculated using the equilibrium constants and activities. Ultimately, the reaction rates within the model are implemented as a function of the difference between the current mass fractions and the equilibrium concentration. The size of the time steps within the process model is controlled globally for the entire system of differential equations by means of the solver (BDF) in use [14]. A more detailed description can be found in the corresponding work by Hay [13]:

$$\dot{m}_{Species\ i} = kd_i \times \left(w_i^{melt} - w_i^{interface}\right) \quad (1)$$

$$K_{eq,i} = e^{\frac{-\Delta G^0}{RT}} = \prod a_i^{v_i} \quad (2)$$

For calculation of the reaction rate as previously described, it is necessary to compute the activity of the species within the liquid melt and slag. Computation of the melt activities is comparably simple, as the melt is largely made of liquid iron with low concentration of dissolved species. In this regard, Hay incorporated both the Wagner interaction parameter formalism (WIPF) [15] and the unified interaction parameter formalism (UIP) [16]. The slag can, however, vary over a wide range. It consists, in part, of metallic oxides (mainly iron oxide) and silica from adhesives of the charged scrap. From a metallurgy point of view, the main requirement on the slag is to promote dephosphorization as well as desulphurization. Both metals are often present in the scrap and coal, yet are undesirable in produced steel [17]. In order to facilitate oxidation of phosphor and sulphur, it is customary to charge basic material such as chalk, lime, or dolomite alongside the scrap in order to raise the activity of oxygen within the slag and lower the activity of oxygen within the liquid melt [18,19]. For this reason, a large portion of the slag is made up of calcium and magnesium oxide from the slag formers. In addition, magnesium oxide is also released into the liquid slag from disintegration of the refractory material. Oversaturation of the slag with CaO and MgO can ultimately lead to the formation of another phase, further increasing the complexity of the system [20]. In addition to the aforementioned species, the slag also contains various sulfides and fluorites to a lesser extent. That being said, due to the high complexity and unavailability of interaction parameters for these species, the EAF model focuses purely on oxides. In Table 1, an exemplary composition of the EAF slag from a steel plant producing structural and engineering steels is listed. Below that, the upper and lower limits (UL and LL) on the mass fractions of each species for the subsequent creation of the data for training of the neural network are shown. The first scenario is intended to match the composition of the slag at the time of tapping, while the second scenario is provided for a more general description. For the calculation of the activities within the slag, the regular solution (RS) model published by Ban-Ya [21] as well as the cell model by Gaye et al. [22] are implemented. Within the RS model, cations are assumed to be randomly distributed within an oxygen–anion matrix. In contrast, the cell model describes the slag in terms of cells composed of a single central anion surrounded by cations [23]. Unfortunately, the interaction parameters published by Ban-Ya are missing information for chromium oxide. These have been partially supplemented by the work of Xiao and Holappa [24]. However, when compared to the results of the commercial software FactSage 6.4 (Version 6.4, GTT Technologies Herzogenrath, Germany) [25], the deviation of both models necessitated the usage of further parameters for correction. Furthermore, while the cell model yields better results than the regular solution model, the computation time is significantly higher, accounting for up to 40% of the EAF model's total runtime. For this reason, estimation of the slag activities using a neural network is investigated. Within this work, the calculation of the slag activity is treated as a pure regression task.

Table 1. Exemplary composition and temperature with upper and lower bounds.

		Temperature	FeO	Cr_2O_3	Al_2O_3	MnO	P_2O_5	SiO_2	MgO	CaO
Unit		K	wt.%							
Slag *			35.6	2.5	6.1	6.4	0.3	11.9	9.6	25.8
Scenario 1: tapping	LL	1700	20	1	2.5	2.5	0	5	5	20
	UL	2000	50	5	10	10	0	15	25	50
Scenario 2: process	LL	1700	0	0	0	0	0	0	0	0
	UL	2000	100	100	100	100	0	100	100	100

* analyzed slag contained further species such as S, TiO_2, and V_2O_5.

The necessary data for training of the model are generated by using the FactSage Equilib module on a dataset generated corresponding to the aforementioned upper and lower bounds. The temperature values are drawn from a uniform distribution. For the composition, two separate scenarios are considered. For the first case the composition range is aimed to match the composition of the slag at tapping. However, depending on the melting rates of the input material, the slag composition varies throughout the process. For this reason, in the second scenario, broader boundaries are chosen, ranging from 0 to 100 percent for each species. Phosphorus oxide is set to zero in both cases, as it is not included in the version 6.4 of the FToxid or FTstel databases used for generation of the training data [26]. In the first scenario, individual mole fractions are drawn from a uniform distribution by applying the stated lower and upper bounds. The composition array is then normalized by division with the overall sum. However, in the second scenario, this method cannot be used. Since all species range from 0 to 100%, by normalizing, all mol fractions are likely to be on a similar level. This way, creating a composition with only one or two major components is virtually impossible, as it would require all other fractions to be very small. To address this issue, the composition array is created in multiple steps. In the first step a single value between 0 and 1 is drawn from a uniform distribution. Subsequently, other $(n-1)$ values are drawn that range from 0 to $(1-x)$ where n denotes the number of species and x denotes the result of the first sample. Using all values, as well as 0 and 1, an array is created. Ultimately, the array is sorted, and the difference between each element and its successor is calculated. The result is assigned to the species at random. This way, the sum of all entries is equal to one, while a single species is likely to have a higher mole fraction than the others. For each scenario, a total of 10,000 compositions is created, 40% of which is used for training of the neural network. In Figure 1a, the generated slag composition of the tapping scenario is shown in the form of a box plot. The slag consists mainly of FeO, CaO, and MgO with lower percentages of the remaining species. Figure 1b also shows the probability distribution of the molar fractions for the second scenario. As can be seen, when drawing from a uniform distribution with subsequent normalization, almost no samples exceed a mole fraction of 0.4. In contrast, when stepwise-creating the composition array, mole fractions of up to 1 can be reached. By nature, this results in a higher overall number of small mole fractions. The composition matrices from scenario 1 and 2 (with stepwise creation) are used in conjunction when training the model to prevent overfitting of the neural network to a specific situation.

For approximation of the chemical activity, a shallow feedforward neural network was chosen. When applying neural nets, most computational effort is during training of the model weights. Predicting the output using the trained model is as simple as matrix multiplication and summation, as such computation of the prediction is not reliant on third-party libraries and can be implemented very efficiently. The network is composed of an input and output layer, as well as a single hidden layer with 80 neurons. The layout of the neural network is depicted in Figure 2a.

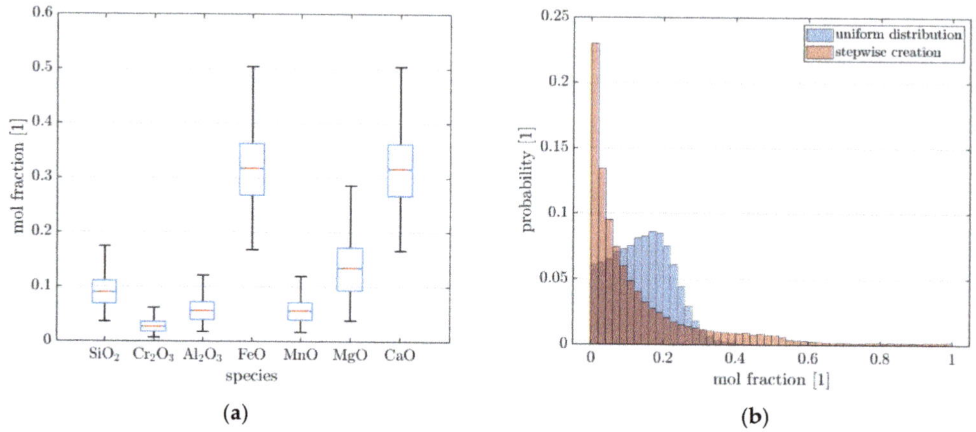

Figure 1. (**a**) Boxplot of generated mole fractions from slag composition at tapping; and (**b**) probability distribution for uniformly drawn and stepwise creation of composition.

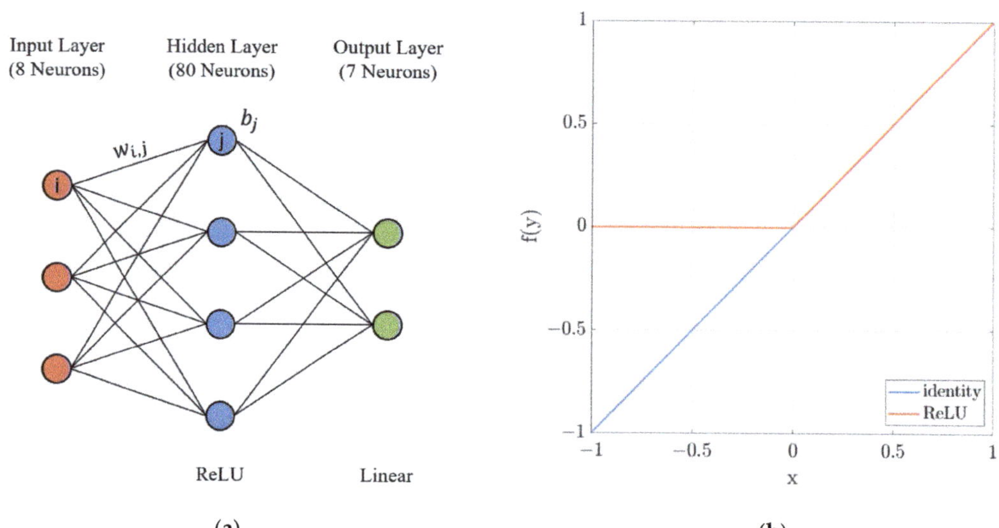

Figure 2. (**a**) Structure of the neural network; and (**b**) response of the activation functions.

As shown in Equation (3), at each node, the input from the previous layer denoted by v_j is first multiplied by a set of weights w_{ij} specific to the neuron i. Afterwards, the corresponding vector is cumulated, and a bias bi is added. The result x is passed onto the next layer by application of an activation function f(x). The output layer utilizes a linear activation function (identity) with a constant value as given by Equation (4), while the neurons within the hidden layer uses a rectified linear unit (ReLU) as stated in Equation (5). Both functions are visualized in Figure 2b. In the recent past, the ReLU has become the default choice for most feedforward networks [27]. While the advantages of the ReLU in terms of the learning rate are neglectable for small networks, it has been chosen in order to introduce a nonlinear transformation to the model. Stacking linear layers, the output would otherwise still be equivalent to a linear combination of the input arguments [28]. The model input consists of the temperature and composition of the slag. In theory, the chemical activity is also dependent on pressure. However, the melt–slag interface is located above the melt, with the pressure inside the EAF being reasonably close to atmospheric pressure. For this reason, the pressure dependency of the activity can be neglected. The output contains the activities of each species at the given temperature and composition.

$$x = \sum_{j=1}^{n} (w_{ij} v_j) + b_i \tag{3}$$

$$f_{linear}(x) = a * x \tag{4}$$

$$f_{ReLU}(x) = \frac{x + |x|}{2} \tag{5}$$

Both the weights and biases of each neuron can be seen as parameters of the neural network. Starting from their initial values, during training, the parameters are adjusted such that the output of the network best matches the reference values. To achieve this, a backpropagation algorithm is applied. Starting from the output layer, the effect of each individual weight on the result is calculated by partial derivation of the obtained overall error. The weights can then be adjusted accordingly before the next iteration [29]. This process is repeated until the output is within a specified tolerance or the maximum number of iterations is reached. In the upcoming section, the results of the surrogate model are compared to the previously used models. In this regard, the results are evaluated based on several metrics. A major aspect is the accuracy of the model prediction. As such, the deviation between the chemical activity calculated by each model and the reference values obtained with FactSage is assessed using the mean absolute error (MAE) as stated in Equation (6). Additionally, the adjusted coefficient of determination (R^2) given in Equation (7) is calculated from the ratio of the residual (RSS) and total sum of squares (TSS). The coefficient is a popular measure for the quality of a fit and is often presented as the percentage of variance of the result which is explained by the linear relationship with the explanatory variables [30]. Finally, the 95% percentile is provided, representing the threshold within which 95% of the results fall. Likewise, 5% of the results have a deviation greater than the threshold. This metric can be interpreted as a worst-case analysis. Due to the dynamic nature of the EAF process, small deviations from the actual result can cancel each other out, while large deviations will significantly affect the overall result of the simulation.

$$MAE = \frac{1}{n} \sum |f_i - y_i| \tag{6}$$

$$R^2 = 1 - \frac{RSS}{TSS} = 1 - \frac{\sum (f_i - y_i)^2}{\sum (y_i - \overline{y})^2} \tag{7}$$

On the other hand, the computation speed of the models is relevant, as the purpose of the surrogate model is not only to closely reproduce the reference value, but to do so in significantly less time than the previous models. In order to eliminate random fluctuations, the execution time is averaged for all samples within the validation set. The calculations are performed in serial, as this resembles the later usage in the EAF model. Furthermore, the computation time of the surrogate model is multiplied by a factor of 8/7 as it is missing the activity of phosphorus oxide. However, it can be argued that the complexity of the neural network remains largely unchanged, as the additional output only affects the size of matrices used for calculation but not the overall structure of the model.

3. Results

In Figure 3, the results of the surrogate model are displayed for MgO, FeO, MnO, and Cr_2O_3. In addition, the activities calculated using the previous models are shown. The solid line represents a perfect fit between the model results and the reference values obtained with FactSage. Values that exceed an activity of 1 are displayed at the corresponding boundary. The performance of the investigated models is also summarized in Table 2 using the previously outlined metrics. As can be seen from both Figure 3 and Table 2, the trained neural network is able to estimate the slag activity more reliably than the RS, as well as the cell model. With the only exception of Cr_2O_3 at activities close to 1, the surrogate model reproduces the FactSage results over the entire range of compositions. This corresponds to an average deviation of 0.016 at maximum and a 95% quantile threshold of 0.065. The coefficient of determination (R^2) ranges from 0.97 to 0.99, showing an exceptional quality of the fit. In contrast, large deviations can be observed for the previous models. This is especially the case for the RS model. The poor representation can be partly attributed to the missing interaction parameters for Cr^{3+} with Fe^{2+}, Mn^{2+}, and P^{5+}. This is confirmed by the substantial deviation for both the mean deviation and 95% percentile of Cr_2O_3. Nonetheless, the results of the RS also show deviations for the remaining species.

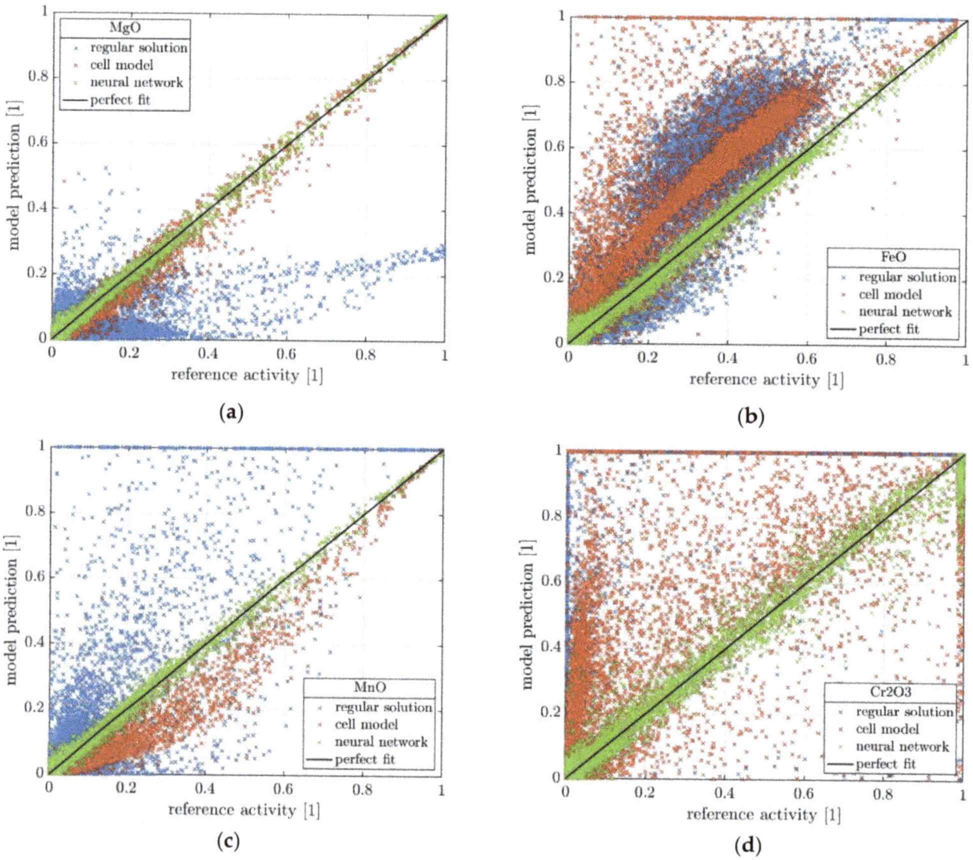

Figure 3. Activities obtained by the cell model and Gaussian process regression compared to the results from FactSage for relevant oxides in the slag: (**a**) magnesium oxide, (**b**) iron oxide, (**c**) manganese oxide, and (**d**) chromium oxide.

Table 2. Overview of the performance of the applied models.

Model	Execution Time for 1000 Samples	Species	MAE	R^2	95% Percentile
RS	0.037 s	SiO_2	0.0279	0.71	0.1483
		Cr_2O_3	0.8059	−10.1	0.9995
		Al_2O_3	0.0089	0.95	0.0563
		FeO	0.1523	−0.09	0.3543
		MnO	0.0651	−0.11	0.2815
		MgO	0.0846	−0.05	0.2424
		CaO	0.0380	−0.09	0.1097
Cell	9.848 s	SiO_2	0.0289	0.65	0.1558
		Cr_2O_3	0.1483	0.15	0.5462
		Al_2O_3	0.0142	0.53	0.0480
		FeO	0.1582	−0.18	0.3823
		MnO	0.0376	0.81	0.1210
		MgO	0.0145	0.97	0.0425
		CaO	0.0058	0.97	0.0191
ANN	0.115 s	SiO_2	0.0086	0.97	0.0379
		Cr_2O_3	0.0163	0.98	0.0655
		Al_2O_3	0.0046	0.99	0.0201
		FeO	0.0112	0.99	0.0354
		MnO	0.0071	0.99	0.0232
		MgO	0.0068	0.99	0.0222
		CaO	0.0041	0.99	0.0137

The RS model performs best for Al_2O_3 and SiO_2, with R^2 scores of 0.95 and 0.71, respectively. Overall, the cell model performs better than the RS model with an R^2 of up to 0.97 for both MgO and CaO. However, the cell model also faces difficulties in determining the activity of FeO and Cr_2O_3. For both species, the activity deviates from the reference by 0.15 on average. In addition, the computational demand of the cell model is roughly 300 times larger when compared to that of the RS model. When running the EAF process model for a heat of 60 min, the ODE solver takes approximately 35,000 calculation steps depending on the smoothness of the furnace operation. By applying the cell model without further simplification, this results in an additional simulation time of over 340 s on the system used for the testing of the model.

In contrast, the neural network takes only about 0.1 s for 1000 iterations (4 s for the entire heat) and is, thus, on a similar scale as the RS model while providing significantly improved results. This comparison neglects the influence of the calculated slag activity on the simulation. In theory, an improved calculation of chemical activities could lead to a more continuous progression of chemical reactions within the model, further reducing the number of evaluation steps required by the solver. However, this is highly dependent on the general operation of the furnace and could not be reliably observed for the process model under investigation. Furthermore, within the process model, additional simplification and optimization is performed, reducing the overall numerical effort.

4. Discussion

Within this work, a shallow feedforward network is applied in order to approximate the chemical activity of the species within a metallurgical slag. The network is used as a surrogate within a comprehensive dynamic process model of the EAF in order to reduce the overall computational demand while maintaining a reasonable accuracy. For the training of the model, a large number of chemical compositions was created to reflect the different stages of the EAF process. The reference activities were then calculated using the commercial software FactSage. Afterwards, the dataset was split into separate datasets for the training and validation of the model. The accuracy of the neural network was compared to the results of the previously applied regular solution and cell model. During validation, it became evident that the neural network outperforms the previous models in terms of the prediction accuracy. At the same time, the computational complexity remained close to that of the regular solution model. This way, by implementing the surrogate model, the runtime of the process model can be minimized when performing, for example, optimization tasks. However, since the modelling approach is not based on physical principles, the neural network should not be used outside the range of the training data. Otherwise, the results may produce unexpectedly large deviations from the actual chemical activities. In such cases, it is advisable to retrain the model or integrate the original FactSage model. While it is possible to directly integrate FactSage's calculation into the EAF process model, this is also associated with an increased overhead and a higher overall runtime.

Within the context of this work, FactSage was used both for the creation of the data used for the training of the surrogate model, as well as for reference when evaluating the results of the investigated models. Therefore, the applied neural network may not necessarily be better suited for calculating the chemical activities. The evaluation of the results only confirms that the model is able to better match the results provided by FactSage. Furthermore, Version 6.4 of the FTstel and FToxid databases are missing data on phosphorus oxide. As noted before, the phosphorus content of the steel is, however, a major concern for ensuring the targeted properties of the produced steel. In summary, the surrogate model is highly dependent on the availability of the underlying reference model and its ability to produce the desired output. That being said, the presented approach can be applied in various contexts and is not limited to the calculation of chemical activities. The benefit of employing a surrogate model in terms of computational effort is, however, dependent on the complexity of the original problem as well as the implementation of the applied methods.

On a side note, the considered EAF process model was implemented using the Python programming language for the reason of its open-source nature and accessibility. As an interpreter language, Python is, however, inherently slower than compiler languages such as C or Fortran [31,32]. Utilizing third-party libraries such as numba, the Python function can be compiled ahead of time [33]. The output of the trained neural network is calculated using basic operations such as matrix multiplication and summation. Therefore, it is not reliant on third-party libraries and can be easily implemented to work in conjunction with numba. This way, the calculation of the slag activity and, by extension, the entire EAF model can by pre-compiled, resulting in a further decrease in runtime.

Author Contributions: Conceptualization, A.R., L.S. and T.-N.E.; methodology, A.R., T.-N.E. and L.S.; software, A.R., T.-N.E. and L.S.; validation, A.R.; resources, T.E.; data curation, A.R. and L.S.; writing—original draft preparation, A.R.; writing—review and editing, A.R., L.S. and T.E.; visualization, A.R.; supervision, T.E.; project administration, T.E.; funding acquisition, T.E. All authors have read and agreed to the published version of the manuscript.

Funding: This research received no external funding.

Data Availability Statement: The original contributions presented in the study are included in the article, further inquiries can be directed to the corresponding author.

Conflicts of Interest: The authors declare no conflicts of interest.

References

1. World Steel Association. *World Steel in Figures: Concise Version*; World Steel Association: Brussels, Belgium, 2023. Available online: https://worldsteel.org (accessed on 6 February 2024).
2. Demus, T.; Reichel, T.; Schulten, M.; Echterhof, T.; Pfeifer, H. Increasing the sustainability of steel production in the electric arc furnace by substituting fossil coal with biochar agglomerates. *Ironmak. Steelmak.* **2016**, *43*, 564–570. [CrossRef]
3. Kirschen, M.; Hay, T.; Echterhof, T. Process Improvements for Direct Reduced Iron Melting in the Electric Arc Furnace with Emphasis on Slag Operation. *Processes* **2021**, *9*, 402. [CrossRef]
4. Hay, T.; Visuri, V.-V.; Aula, M.; Echterhof, T. A Review of Mathematical Process Models for the Electric Arc Furnace Process. *Steel Res. Int.* **2021**, *92*, 2000395. [CrossRef]
5. Carlsson, L.S.; Samuelsson, P.B.; Jönsson, P.G. Predicting the Electrical Energy Consumption of Electric Arc Furnaces Using Statistical Modeling. *Metals* **2019**, *9*, 959. [CrossRef]
6. Meier, T. *Modellierung und Simulation des Elektrolichtbogenofens*; Verlagshaus Mainz: Mainz, Germany, 2016; ISBN 978-3-95886-129-9.
7. Logar, V.; Dovžan, D.; Škrjanc, I. Modeling and Validation of an Electric Arc Furnace: Part 1, Heat and Mass Transfer. *ISIJ Int.* **2012**, *52*, 402–412. [CrossRef]
8. Logar, V.; Dovžan, D.; Škrjanc, I. Modeling and Validation of an Electric Arc Furnace: Part 2, Thermo-chemistry. *ISIJ Int.* **2012**, *52*, 413–423. [CrossRef]
9. Pfeifer, H.; Nacke, B.; Beneke, F. *Praxishandbuch Thermoprozesstechnik: Band II: Anlagen-Komponenten-Sicherheit*, 3rd ed.; Vulkan: Essen, Germany, 2022; ISBN 9783802731228.
10. Ben Bowman, K.K. *Arc Furnace Physics*, 1st ed.; Stahleisen: Düsseldorf, Germany, 2009; ISBN 978-3514007680.
11. Graham, K.J. Integrated Ladle Metallurgy Control. Ph.D. Thesis, McMaster University, Hamilton, ON, Canada, 2008.
12. MacRosty, R.D.M.; Swartz, C.L.E. Dynamic Modeling of an Industrial Electric Arc Furnace. *Ind. Eng. Chem. Res.* **2005**, *44*, 8067–8083. [CrossRef]
13. Hay, T.; Reimann, A.; Echterhof, T. Improving the Modeling of Slag and Steel Bath Chemistry in an Electric Arc Furnace Process Model. *Metall. Mater. Trans. B* **2019**, *50*, 2377–2388. [CrossRef]
14. Shampine, L.F.; Reichelt, M.W. The MATLAB ODE Suite. *SIAM J. Sci. Comput.* **1997**, *18*, 1–22. [CrossRef]
15. Lupis, C.; Elliott, J. Generalized interaction coefficients: Part II: Free energy terms and the quasi-chemical theory. *Acta Metall.* **1966**, *14*, 1019–1032. [CrossRef]
16. Pelton, A.D.; Bale, C.W. A modified interaction parameter formalism for non-dilute solutions. *Metall. Trans. A* **1986**, *17*, 1211–1215. [CrossRef]
17. Basu, S.; Seetharaman, S.; Lahiri, A.K. Thermodynamics of Phosphorus and Sulphur Removal during Basic Oxygen Steelmaking. *Steel Res. Int.* **2010**, *81*, 932–939. [CrossRef]
18. Michalek, K.; Čamek, L.; Gryc, K.; Tkadlečková, M.; Huczala, T.; Troszok, V. Desulphurization of the high-alloy and middle-alloy steels under the conditions of an eaf by means of synthetic slag based on CaO-Al_2O_3. *Mater. Technol.* **2012**, *46*, 297–303.
19. Schrama, F.N.H.; Beunder, E.M.; van den Berg, B.; Yang, Y.; Boom, R. Sulphur removal in ironmaking and oxygen steelmaking. *Ironmak. Steelmak.* **2017**, *44*, 333–343. [CrossRef]
20. Kirschen, M. Visualization of Slag Data for Efficient Monitoring and Improvement of Steelmaking Slag Operation in Electric Arc Furnaces, with a Focus on MgO Saturation. *Metals* **2021**, *11*, 17. [CrossRef]
21. Ban-Ya, S. Mathematical Expression of Slag-Metal Reactions in Steelmaking Process by Quadratic Formalism Based on the Regular Solution Model. *ISIJ Int.* **1993**, *33*, 2–11. [CrossRef]
22. Gaye, H.; Lehmann, J.; Matsumiya, T.; Yamada, W. A statistical thermodynamics model of slags: Applications to systems containing S, F, P_2O_5 and Cr oxides. In Proceedings of the 4th International Conference on Molten Slags and Fluxes, Sendai, Japan, 8–11 June 1992; ISIJ: Sendai, Japan, 1992; pp. 103–108.
23. Hay, T. Mathematische Modellierung des Elektrostahlverfahrens. Ph.D. Thesis, RWTH Aachen University, Aachen, Germany, 2021.
24. Xiao, Y.; Holappa, L. Thermodynamics of slags containing chromium oxides. In Proceedings of the INFACON 7, Trondheim, Norway, 11–14 June 1995; Norwegian Ferroalloy Research Organization (FFF): Trondheim, Norway, 1995.
25. Bale, C.W.; Bélisle, E.; Chartrand, P.; Decterov, S.A.; Eriksson, G.; Gheribi, A.E.; Hack, K.; Jung, I.-H.; Kang, Y.-B.; Melançon, J.; et al. FactSage thermochemical software and databases, 2010–2016. *Calphad* **2016**, *54*, 35–53. [CrossRef]
26. GTT, C. FactSage Database Overview. Available online: https://www.factsage.com/facthelp/FSNew_Databases.htm (accessed on 31 January 2024).
27. Goodfellow, I.; Bengio, Y.; Courville, A. *Deep Learning*; MIT Press: Cambridge, MA, USA, 2016.
28. Szandała, T. Review and Comparison of Commonly Used Activation Functions for Deep Neural Networks. In *Bio-Inspired Neurocomputing*; Bhoi, A., Mallick, P., Liu, C.M., Balas, V., Eds.; Springer: Singapore, 2021; pp. 203–224. ISBN 978-981-15-5495-7.
29. Rumelhart, D.E.; Hinton, G.E.; Williams, R.J. Learning representations by back-propagating errors. *Nature* **1986**, *323*, 533–536. [CrossRef]
30. Renaud, O.; Victoria-Feser, M.-P. A robust coefficient of determination for regression. *J. Stat. Plan. Inference* **2010**, *140*, 1852–1862. [CrossRef]

31. Romer, T.H.; Lee, D.; Voelker, G.M.; Wolman, A.; Wong, W.A.; Baer, J.-L.; Bershad, B.N.; Levy, H.M. The structure and performance of interpreters. In Proceedings of the Seventh International Conference on Architectural Support for Programming Languages and Operating Systems, Cambridge, MA, USA, 1–4 October 1996; Association for Computing Machinery: New York, NY, USA, 1996; pp. 150–159, ISBN 0897917677.
32. Barany, G. Python Interpreter Performance Deconstructed. In Proceedings of the Workshop on Dynamic Languages and Applications, Edinburgh, UK, 12 June 2014; Association for Computing Machinery: New York, NY, USA, 2014; pp. 1–9, ISBN 9781450329163.
33. Anaconda. Numba: User Manual. Available online: https://numba.readthedocs.io/en/stable/user/index.html (accessed on 3 June 2024).

metals

MDPI

Article

Consequences of the Direct Reduction and Electric Steelmaking Grid Creation on the Italian Steel Sector

Gianluca Dall'Osto *, Davide Mombelli and Carlo Mapelli

Dipartimento di Meccanica, Politecnico di Milano, Via La Masa 1, 20156 Milano, Italy; davide.mombelli@polimi.it (D.M.); carlo.mapelli@polimi.it (C.M.)
* Correspondence: gianluca.dallosto@polimi.it

Abstract: The consequences on the Italian steel sector following the conversion of the sole integrated steel plant and the establishment of a direct reduction/electric arc furnace (DR/EAF) grid in the period 2022–2050 were analyzed. Imported natural gas (pathway 0), green hydrogen (pathway 1) and biomethane (pathway 2) were studied as possible reducing gases to be exploited in the DR plant and to be introduced as a methane substitute in EAFs. The results showed that the environmental targets for the sustainable development scenario could be achieved in both 2030 and 2050. In particular, the main reduction would occur by 2030 as a result of the cease of the integrated plant itself, allowing for an overall reduction of 71% of the CO_2 emitted in 2022. On the other hand, reaching the maximum production capacity of the DR plants by 2050 (6 Mton) would result in final emission reductions of 25%, 80% and 35% for pathways 0, 1 and 2, respectively. Finally, the creation of a DR/EAF grid would increase the energy demand burden, especially for pathway 1, which would require three times as much green energy as pathway 0 and/or 2 (36 TWh/y vs. ca. 12 TWh/y).

Keywords: CO_2 emissions; iron and steel industry; direct reduction; decarbonization; green H_2; biomethane

1. Introduction

Steel, along with cement, plastics and ammonia, has always been recognized as one of the most crucial products for the industrial sustenance of both advanced and developing economies. Of the 1884 Mton of crude steel produced in 2022, 71.5% was obtained through the integrated cycle, while only 28.2% was covered by the scrap recycling route, a share of production that has remained more than constant over the past 20 years despite the long-anticipated desire and need for its increase [1].

Indeed, despite the growing demand for steel products, the parallel growth of social awareness regarding the high environmental impact associated with steel production has raised shadows about the future of this industry. According to the IEA latest measurements and data collection [2], the steel sector is responsible for 2.8 Gton of CO_2 emissions per year, accounting for 8% of total energy system emissions, a value that raises to 10% if indirect emissions from electricity generation are included.

To achieve the climate goals of the Paris Agreement and the Sustainable Development Scenario (SDS) for the steel and iron sectors, direct emissions would have to drop dramatically to 0.6 tons of CO_2 per ton of crude steel, corresponding to 1.2 Gton of CO_2 emitted per year, more than half of the current value [3,4]. Therefore, to meet this need, the replacement of fossil fuel with alternative carbon sources [5], the use of low-or-zero emission gaseous streams instead of natural gas (NG) [6], the exploitation of carbon capture and storage (CCS) [7], the continuous technological upgrade of existing steel plants as well as the creation of steel plants coupled with green hydrogen (H_2) direct reduction (H_2-DR) [8] or electrowinning (EWIN) [9] are generally considered promising pathways [10]. Consequently, several breakthrough projects have been created in the last decade to investigate the feasibility of one or more specific technologies for the carbon footprint mitigation [11].

Despite the large number of projects, the actual implementation of which is underway or planned, the scrap recycling route is overall considered to be the production route that will become the predominant one in the near future. In fact, it is predicted that, by 2050, the actual production of steel by melting scrap inside an electric arc furnace (EAF) will almost double its share, reaching a value of 53% [12]. The main reason for this trend can be easily identified in the specific CO_2 emission per ton of crude steel of the EAF compared to that of the integrated cycle, based on the reduction of iron ore by a blast furnace (BF) and basic oxygen furnace (BOF), namely 10^2 kg_{CO2eq}/t_{CS} vs. 10^3 kg_{CO2eq}/t_{CS} [13]. It is noteworthy that, even in the context of the EAF, several studies have focused on the process decarbonization mainly through the introduction of alternative carbon sources, either as a reductant or a foaming agent (Table 1).

Table 1. Main usages and studies of alternative carbon sources in the scrap recycling route.

Scale Investigated	Main Usage/Scope of the Research	Material Used	Reference
Laboratory	Foaming agent	Biochar	[14]
Laboratory	Foaming agent	Biochar	[15]
Laboratory, pilot and industrial	Overall feasibility as carbon substitute	Biochar and hydrochar	[16]
Laboratory and industrial	Carburizing agent	Biochar	[17]
Simulation and industrial	Overall feasibility as carbon substitute	Biochar	[18]
Simulation of an industrial EAF	Foaming and carburizing	Biochar and hydrochar	[19]
Industrial	Overall feasibility as carbon substitute	Biochar and hydrochar	[20]

Along with the transition to a more scrap-based steel production, the demand for clean iron sources to be introduced as charging material inside the EAF is expected to increase to counter the inevitable depletion of high-quality scrap in the near future [21]. Specifically, with clean iron sources, the intent is a pollutant-free iron obtained by the direct reduction (DR) of iron ores to obtain so-called sponge iron, or in other words, direct reduced iron (DRI) or hot briquetted iron (HBI).

The production of these sources of clean iron is not new to the steel industry, and several commercial plants (e.g., MIDREX® (Charlotte, NC, USA), Rotary Kiln (FLSmidth, Inc., Tucson, AZ, USA) and HyL (Tucson, AZ, USA)/Energiron® (Buttrio, Italy)) are active around the world with the main producers belonging to the Middle East/North Africa and Asia/Oceania regions. In particular, India is currently leading DRI production with 43.55 Mton out of the total 127.36 Mton produced in 2022 alone [21]. According to market trends and company estimates, the largest increase in the mid-term will occur in the European Union (EU) with a dramatic increase in production capacity of 50 times that of today (20 Mton in 2030 compared to 400 kton in 2019). NG is expected to remain the main reducer, although green H_2 will account for a significant share of production mainly in Scandinavia and perhaps other regions [22]. In addition, biosyngas, which can be obtained through the gasification of biomass within fluidized bed gasifiers, has reached a mature technological level (TRL 7–9) and could play an important role in DRI production as an alternative reducer to both NG and green H_2 [23].

It will, therefore, be necessary for most countries that rely or will rely on the scrap recycling route to decide whether it is more cost-effective to build a DR/EAF grid within national borders or, as is currently done, to rely on importing DRI. In this regard, Italy, currently the world's 11th largest producer, can be considered a worthy case study as its steel production is based almost exclusively on the scrap recycling route (18 Mton produced by EAF vs. 3 Mton by BF/BOF in 2022), and as there is still no DR plant within the national borders, it relies on the import of approximately 2 Mton of clean iron sources (e.g., DRI, HBI and pig iron) per year and approximately 5 Mton of scrap per year to feed its 37 EAF production sites [24]. To make up for this deficiency, a memorandum was signed in 2023 for the decarbonization of the Italian steel sector with a focus on the conversion of the BF/BOF plant of Taranto into a DR plant [25,26]. In order to analytically understand the feasibility of this transition and which strategies/technologies are best suited to actually achieve it, techno-economic or techno-assessment modeling appears to be an appropriate methodology. In fact, they have proven to be a more than reliable tool for highlighting the main barriers that the transition from hard-to-abate to near-zero emissions term would entail, especially if the analysis is limited to individual countries or technologies. In the EU context, Mandova et al. [27] focused their attention on the implementation of biomass and bio-CCS technologies to primary steelmaking, highlighting an overall CO_2 reduction of 20% and 50%, respectively. Still in the primary steelmaking framework, Fischedick et al. [28] as well as Arens et al. [29] studied the possible future of the German steel industry. They both concluded that the high production share of the integrated cycle (approximately 70%) created an urgent need for emissions cutting in the short to medium term achievable through the use of heat-recovery technologies, an increase in the scrap recycling production share and the use of by-products for the production of base chemicals. In Norway, Bhaskar et al. [30] developed a techno-economic assessment of a H_2-DR/EAF plant, highlighting that, though on the one hand the availability of green electricity and magnetite ore is the main influencing factor, on the other the national offshore wind energy potential makes the country an ideal location for hydrogen-based steel production. A similar pathway was suggested in Sweden by Toktarova et al. [31], who concluded that the H_2-DR/EAF process could abate 10% the total Swedish CO_2 emissions, but the main challenge would be the 14 TWh demanded for the electrification of the primary steel production while maintaining constant steel production throughout the time period of the analysis. Furthermore, they also highlighted that the implementation of CCS and biomass in the BF along with the transition of the primary steel production plant to the scrap recycling route would result in an 80% reduction with respect to the current Sweden steel sector emissions. Furthermore, CCS technologies must also overcome the obstacles caused by the removal of contaminants (e.g., N_2, SO_2, NO_x, fly ash, trace metals) in the furnace flue gases before the actual capture of CO_2 that would otherwise hinder the efficiency of the process [32].

Despite the importance of the steel sector in Italy, a specific technical assessment model on the feasibility of creating a DR/EAF grid is still lacking. To date and to the authors' knowledge, there are only two papers that investigated the sector from a general point of view, and they were published in 2016 and 2022 by Renzulli et al. [33] and Mapelli et al. [13], respectively. However, the former focuses exclusively on the environmental impact of the ILVA plant, while the latter evaluates possible options for impact mitigation, focusing on the advantages and disadvantages they would bring to each of the specific steel production pathways while leaving out the variation over time.

Therefore, this study aims to investigate the transition of the Italian steel sector toward the creation of a national DR/EAF grid with the goal of filling the lack of a technology assessment, which is currently present in the literature, while also providing new insights into the consequences of this transition. The focus is on the 2030 and 2050 CO_2 emission thresholds and the weight that green H_2-DR would have on renewable electricity and bio-CH_4-DR would have on biomethane production. In addition, the results will be compared with those of a traditional NG-DR plant in order to highlight the barriers and risks associated with such a transition in the near future.

2. Materials and Methods

The proposed analysis derives and estimates the amount of CO_2 and electricity demand associated with the creation of a DR/EAF grid in the Italian context over a time period from 2022 to 2050. Prior to the analysis, the following assumptions were made:

- A constant steel production of 22 Mton is assumed between 2022 and 2050. According to Federacciai [24], Italian steel production has fluctuated between 31 Mton in 2007 and 21 Mton in 2022, mainly due to the macro-economic context. On the other hand, if only the time period between 2015 and 2022 is considered, the overall production has remained more constant with an average output of approximately 22 Mton.
- The share of the scrap recycling route is assumed to increase linearly until 2030, the year in which the production relies solely on electric furnaces. After 2030, it is assumed

that the amount of EAFs that will begin to exploit domestically produced DRI and the annual capacity of the DR plants will increase at a rate equal to that described by Arens et al. [29].

- The metallic charge of EAF is assumed to be covered for 10 wt.% by DRI as the only source of clean iron, whereas the import of scraps and DRI is assumed to be ensured for the entire time period of the analysis [34].

2.1. Model Input Parameters and Pathways

The amount of CO_2 and electricity demand related to the Italian steel industry were evaluated based on the results obtained by Mapelli et al. [13], which are summarized in Table 2. The evaluation considered three DR/EAF pathways to show the differences and advantages of DRI production via NG (pathway 0), green H_2 (pathway 1) produced by water electrolysis and exploiting photovoltaic (PV) renewable energy and, finally, biomethane (pathway 2). Furthermore, in pathways 1 and 2, it is assumed that the EAFs that will exploit internally produced DRI will replace methane with the corresponding reducing gas in the pathway. Finally, steel casting, hot and cold working and coating are not included in all pathways because of their significantly lower energy consumption and carbon emissions compared to production.

Table 2. Specific CO_2 emissions and energy demand per ton of steel per type of production process and pathway [13].

Pathway, Process	Specific CO_2 Emissions, ton_{CO2}/ton_{steel}	Specific Energy Demand, kWh/ton_{steel}
Pathway 0		
Blast furnace with basic oxygen furnace (BF/BOF)	2.515	131.25
Electric arc furnace (EAF)	0.135	514
Direct reduction plant fed by natural gas (DR)	0.77	123.5
Pathway 1		
Blast furnace with basic oxygen furnace (BF/BOF)	2.515	131.25
Electric arc furnace (EAF)	0.135	514
EAF fed by green H_2 (H_2-EAF)	0.092	510
Green H_2 production for EAF	0	379
Direct reduction plant fed by green H_2 (H_2-DR)	0	123.5
Green H_2 production for DR	0	2552
Pathway 2		
Blast furnace with basic oxygen furnace (BF/BOF)	2.515	131.25
Electric arc furnace (EAF)	0.135	514
EAF fed by biomethane (Bio-CH_4-EAF)	0.12	514
Biomethane production for EAF	0.018	2.2
Direct reduction plant fed by biomethane (Bio-CH_4-DR)	0.32	123.5
Biomethane production for DR	0.28	34.7

2.2. Key Mathematical Equations

The annual CO_2 emissions for each of the three pathways were evaluated by considering the specific emissions of the production cycle, their share and the total amount of steel and DRI produced per year.

$$TE_{i,t} = \left(\sum_j s_{j,i} \cdot \chi_{j,i,t} + s_{DR,i} \cdot \xi_{i,t} \right) \cdot P_t \quad (1)$$

where *TE* is the total CO_2 emission expressed in Mton, *s* is the specific CO_2 emission related to steel production and reducing gas production expressed in ton_{CO2}/ton_{steel}, χ is the share of the specific production cycle, ξ is the parameter that considers the amount of production capacity of the DR plant per year, and *P* is the amount of steel produced per year expressed in Mton. The symbols *j*, *i* and *t* represent the specific production cycle (integrated cycle, scrap recycling and DR), the pathway and the year.

Similarly, the annual energy demand for each of the three pathways was evaluated by considering the specific emissions of the production cycle, their share and the amount of DRI introduced inside the furnaces and the total amount of steel produced per year.

$$TEC_{i,t} = \left(\sum_j q_{j,i} \cdot \chi_{j,i,t} + q_{DR,i} \cdot \xi_{i,t} \right) \cdot P_t \quad (2)$$

where *TEC* is the total electrical consumption expressed in TWh, *q* is the specific energy demand related to steel production and reducing gas production expressed in TWh/ton_{steel}, χ is the share of the specific production cycle, ξ is the parameter that considers the amount of production capacity of the DR plant per year, and *P* is the amount of steel produced per year expressed in Mton. The symbols *j*, *i* and *t* represent the specific production cycle (integrated cycle, scrap recycling and DR), the pathway and the year.

2.3. Sensitivity Analysis

Sensitivity analysis was performed on the total CO_2 emissions to highlight the level of decarbonization achievable in the three DR/EAF pathways by applying Equation (1) for years 2030 and 2050. The comparison was performed with respect to the IEA projection for the SDS of the corresponding year per ton of steel produced [12].

Furthermore, it is strongly discussed that, in addition to direct CO_2 emissions, the decarbonization of steel and, particularly, electrical steelmaking are highly dependent on the level of decarbonization of the national power grid; hence, a sensitivity analysis was performed on the energy demand of the steel production, evaluated by applying Equation (2) with respect to the renewable energy expected in 2030 and 2050. To date in Italy, more than 70% of the energy is imported with most of it of fossil origin. Hence, to decrease this dependency, the Integrated National Plan for Energy and Environment set very ambitious targets for renewables, aiming to reach 30% in total energy consumption and 55% in electricity generation in 2030. It is foreseen that its pursuit will increase the photovoltaics PV capacity to 52 GW in 2030 and, in order to achieve the carbon neutrality imposed by the EU, to 200 GW in 2050 [12,35,36]. Based on the latest data provided by Terna [37], the resulting TWh would account for approximately 57 TWh in 2030 and 218 TWh in 2050.

Finally, for pathway 2, a sensitivity analysis was performed on the amount of biomethane required compared to the total potential production from biomass. Specifically, biomethane production is assumed to reach a total potential of approximately 6 Gm^3 in 2030 and 13.5 Gm^3 in 2050, as estimated by the European Biogas Association [38,39].

3. Results

3.1. Outline of the Steel Production

Figure 1a shows the amount of steel produced by process type and that of domestically produced DRI (DRI-IT) regardless of the reducing agent assumed in the specific pathway.

Following a conservative approach regarding DRI production, the electrical steelmaking was divided into two separate routes (EAF and DRI-IT/EAF). Specifically, the orange area represents the portion of EAF plants that will continue to take advantage of imported DRI, while the blue area represents the electric steel plants that will feed furnaces with DRI-IT and replace methane with green H_2 (pathway 1) or biomethane (pathway 2). This is in order to avoid an over-demand for DR plant production in the initial years, allowing for them to reach their full production potential without compromising DRI quality.

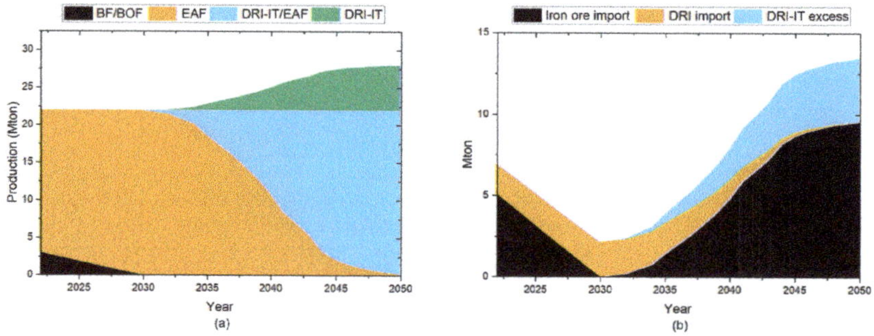

Figure 1. (**a**) Steel and DRI production by process (BF/BOF: integrated cycle, EAF: scrap recycling route, DRI-IT/EAF: scrap recycling route utilizing domestic produced DRI, DRI-IT: domestic produced DRI), (**b**) required iron ore, DRI import and excess of DRI-IT with respect to that charged into DRI-IT/EAF available for the Italian steel sector from 2022 to 2050.

To highlight the consequences of setting up the DR plants, Figure 1b shows the changes in the import demand for DRI and iron ore as well as the amount of excess DRI-IT available that could be further introduced into the furnaces or exported. The net difference between DRI-IT charged and DRI-IT produced, depicted in Figure 1b, was estimated by considering an EAF metal charge covered for 10 wt.% by DRI [34].

Although the production of DRI-IT would be able to cover the demand of the furnaces while also producing an excess of material for each year of the analysis, the main consequence of the domestic DRI production would be an increase in the import of iron ores. Indeed, an amount equal to that currently imported for the BF/BOF plant (5 Mton) would again be required in 2040 and would almost double in 2050, assuming the best-case scenario and the maximum yield of iron ore reduction in which approximately 1.55 ton of iron ore (64 wt.% Fe) is needed for 1 ton of DRI [40].

Remarkably, if the excess of DRI-IT is reintroduced directly into the domestic market, the import of DRI can be abandoned as early as 2038, the year when DR plants reach a level of maturity and productivity that can fully cover furnace demand. Furthermore, also in the same year, the amount of imported DRI would be less than the 5 Mton in 2022 and equal to 3.2 Mton.

3.2. *Estimation of the Pathways' CO_2 Emissions*

Figure 2 illustrates the CO_2 related to each pathway and unbundled by type of production process, either intended as steel production or the production of gaseous streams to be introduced within the furnaces or DR plant.

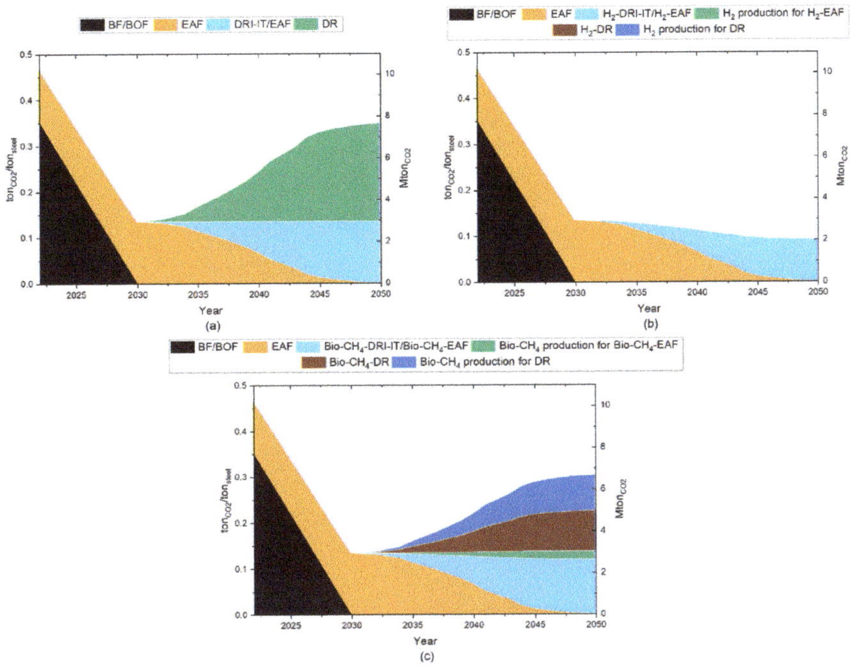

Figure 2. Specific and total CO_2 intensities for (**a**) pathway 0 (use of NG), (**b**) pathway 1 (use of green H_2) and (**c**) pathway 2 (use of biomethane).

In each pathway, the largest reduction in total CO_2 emissions is expected between 2022 and 2030 due to the closure of the BF/BOF plant and its replacement by electric steelmaking. Specifically, by 2030, emissions would be reduced by 71% compared to 2022 while maintaining a constant steel production of 22 Mton. In fact, of the 10.30 $Mton_{CO2}$ emitted in 2022, the BF/BOF plant alone accounts for 7.75 $Mton_{CO2}$. The value appears more than abnormous when compared with that of the EAF plants for the same year (2.55 $Mton_{CO2}$) and bearing in mind that only 14% of production was covered by the integrated cycle.

Since pathway 0 (Figure 2a) assumes a continuity in the use of methane in the EAFs, the value of total emissions from electric steelmaking will remain constant over the period 2030–2050 (2.97 $Mton_{CO2}$). On the contrary, the start-up of NG-DR plants from 2030 onward would lead to an increase in the sector emissions until 2050 due to the reaching of the DR plants' full capacity (6 $Mton_{DRI\text{-}IT}$). Specifically, DRI-IT production would be responsible for 60% of the total 7.67 tons of CO_2 emitted by the sector in 2050, which, though it would still be less than that emitted by the 3 Mton of steel produced by the BF/BOF plant in 2022 (7.75 $Mton_{CO2}$), would not provide a significant effective decarbonization. On the other hand, in 2038, the year when it might be possible to cover the desired 10% of the metal charge of all EAFs with DRI-IT, the value of emissions associated with DR plants would account for 35% of the total 4.58 Mton of CO_2 emitted by the sector, 1.7 times lower than that associated with DRI-IT production in 2050.

The lowest CO_2 emissions are achievable through the application of pathway 1 (Figure 2b) through the substitution of methane with green H_2 in electric steelmaking and the charging within the furnaces of zero-emission H_2-DRI-IT (2.03 $Mton_{CO2}$). Specifically, in 2050, emissions would be reduced by 80% compared to the 2022 value but only 30% compared to the 2030 value. On the other hand, in 2040, when steel production is equally divided between traditional EAF and H_2-DRI-IT/H_2-EAF, the total emissions would be comparable to those of electric steelmaking in 2022 (2.50 $Mton_{CO2}$ vs. 2.55 $Mton_{CO2}$), which can be considered a more than positive result when considering that scrap production would increase by 3 Mton in the same time period.

Finally, pathway 2 (Figure 2c) shows an intermediate situation in which total emissions could be reduced by 3.6 $Mton_{CO2}$ in 2050, representing a 35% mitigation compared to 2022, mainly due to emissions related to Bio-CH_4-DR plants and their respective biomethane production (1.95 $Mton_{CO2}$ vs. 1.70 $Mton_{CO2}$). It is noteworthy to highlight that, if only emissions related to steel production are considered, the decrease would be 74% over the same time period, which rises to 70% if those related to the production of biomethane to be introduced into the furnaces are also considered. Finally, contrary to pathway 1, if the comparison is limited to electric steelmaking in the periods 2022–2040 and 2022–2050, the emissions would increase by 0.02 $Mton_{CO2}$ and 0.06 $Mton_{CO2}$, respectively.

3.3. *Estimation of the Pathways' Energy Demand*

Figure 3 shows the energy demand of each route broken down by type of production cycle and production of gaseous streams to be fed into the furnaces or DR plants.

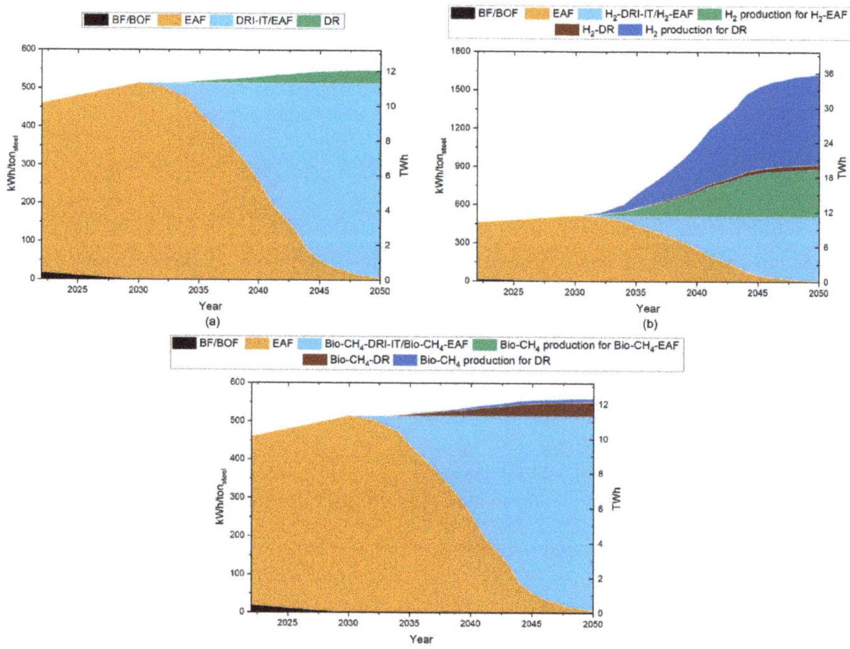

Figure 3. Specific and total energy demand for (**a**) pathway 0 (use of NG), (**b**) pathway 1 (use of green H_2) and (**c**) pathway 2 (use of biomethane). Note the different scale of the y-axis of (**c**).

The results indicate that none of the routes studied are able to achieve energy savings over the entire period examined. Specifically, in the period 2022–2030, the supplanting of the BF/BOF plant with electric steelmaking has, as its main consequence, a 12% increase in the sector energy demand (10.13 TWh vs. 11.31 TWh), which increases further in the following 20 years due to the startup of DR plants.

Pathway 0 (Figure 3a) achieved the lowest increase in energy demand due to the import of NG gas with the sector aggravation on the power grid compared to 2022 being 15% and 19% in 2040 and 2050, respectively. Focusing only on the DR plants in 2040, the year when DRI-IT production would be equal to that of the crude steel produced by the BF/BOF plant in 2022 (3 Mton), the results showed a slight energy savings of 0.02 TWh.

In contrast, at full capacity of the DR plants in 2050, their demand would reach 0.75 TWh, 1.8 times that of the BF/BOF plant in 2022.

The highest energy demand is projected for pathway 1 (Figure 3b) due to the abnormous burden of H_2 production for both EAF and DR plants on the national power grid, which would account for 23.81 TWh of the total 35.79 TWh required by the sector in 2050. Even when considering hydrogen electric steelmaking alone, the demand would be higher than pathway 0 for any year analyzed, as the production of green H_2 to be used as a substitute for NG within electric furnaces would account for approximately 40% of the total demand of electric steelmaking, thus almost doubling the final value of the scrap recycling route. Therefore, it is important to consider the demand in 2038, when the energy demand settles at 19 TWh due to the lower production share of H_2-EAF compared to traditional EAF (34% vs. 66%) and H_2-DRI-IT production (2 $Mton_{DRI\text{-}IT}$).

Finally, pathway 2 (Figure 3b) appears to be an almost identical scenario to pathway 0 due to the low energy demand required for biomethane production. Specifically, an increase in energy demand of 2.19 TWh is expected in 2050 compared to 2022, which is more than comparable to that observed for pathway 0 over the same time period (1.93 TWh). Similarly, the observation made for pathway 0 in 2038 and 2040 could be translated to pathway 1 due to the negligible increase in energy demand (+0.10 TWh and +0.15 TWh compared to pathway 0 in 2038 and 2040).

3.4. Sensitivity Analysis

The estimated CO_2 emissions and energy demand were verified by comparing them with the SDS target provided by the IEA [12] and the national PV energy availability projected in the National Integrated Energy and Environment Plan [35–37] in 2030 and 2050, respectively. The estimated results from this study are given in Table 3.

Table 3. Estimated specific CO_2 emission and energy demand in 2030 and 2050 by pathway. Comparison with IEA target for SDS and Italian PV energy.

Year	Pathway	Specific CO_2 Emissions, ton_{CO2}/ton_{steel}		Energy, TWh	
		Estimated	Target [12]	Estimated Demand	Target PV Capacity [35–37]
2030	0	0.13	1.2	11.31	57
	1	0.13		11.31	
	2	0.13		11.31	
2050	0	0.34	0.60	12.06	218
	1	0.09		35.79	
	2	0.30		12.32	

For each pathway, the level of specific CO_2 emissions in 2030 is approximately 10 times lower than the SDS target (0.13 ton_{CO2}/ton_{steel} vs. 1.2 ton_{CO2}/ton_{steel}). Similarly, though the use of NG and biomethane in pathways 0 and 2 results in an increase in specific emissions over time, both values in 2050 are approximately half the strict SDS values imposed in the same year (0.34 ton_{CO2}/ton_{steel} and 0.30 ton_{CO2}/ton_{steel} vs. 0.60 ton_{CO2}/ton_{steel}).

The direct comparison with PV target capacities showed that, in principle, there would be enough renewable energy to be allocated to the full electric steel industry in Italy in 2030 and to cover the additional demand of DRI-IT production in 2050, regardless of the gas stream used. Specifically, in the former case, the steel sector would burden 19.84% of the PV renewable energy grid, and in the latter case, 5.53%, 16.41% and 5.65% based on the selected gaseous stream, imported NG and domestically produced green H_2 or biomethane, respectively. Finally, the amount of biomethane needed for DRI-IT production and as a substitute for methane within the EAFs would weigh 21.70% of the projected potential to 2050 (Table 4).

Table 4. Estimated biomethane demand in 2050 by pathway 2. Comparison with foreseen data by European Biogas Association.

Year	Pathway	Biomethane, Gm^3	
		Estimated	Expected [39]
2050	2	2.93	13.5

4. Discussion

One of the objectives of this study was to explore how three different pathways, based on NG, green H_2 and biomethane as the main gas stream, could change the emissions and energy demand of the Italian steel sector during the establishment of a DR/EAF grid.

It should be emphasized that, because of the uncertainty about the future of the BF/BOF plant in Taranto, this study should be understood as explanatory and not as an actual projection [41,42]. Furthermore, although this study does not focus primarily on the change in domestic and imported scrap, it is clear that, due to the replacement of the BF/BOF plant with EAF plants, an increase should be expected. As discussed by Pauliuk et al. [43], global scrap availability is expected to increase due to stockpiling in emerging economies, while availability in the EU is expected to stabilize. Thus, the main problem will be the decrease in high-quality scrap to be fed into the EAFs and the consequent increase in demand for clean iron sources for European electric steel plants. Consequently, it should be kept in mind that the expected increase in DRI/HBI production will inevitably generate additional problems related to the demand for high-quality iron ore, its availability and the resulting accelerated depletion of mineral resources. Indeed, if, on the one hand, the results showed that the sole Italian integrated plant conversion to a DR plant would be able to produce an amount of DRI that can cover up to 28% of the national EAF metallic charge by 2050, then, on the other hand, the amount of imported iron ore will be almost doubled compared to the current amount, regardless of the route taken.

Regarding the environmental impact and energy demand, the highest benefits were observed for the closure of the BF/BOF plant in 2030 (−7.33 $Mton_{CO2}$ emitted and +1.11 TWh of energy demand with respect to the 2022 value). Pathway 1 alone can slightly reduce overall emissions by another 1 $Mton_{CO2}$ by 2050 at the expense of a significantly greater burden on the renewable energy grid. The creation of a national H_2-DR/EAF grid is thus bound to the achievement of a PV capacity growth rate twice the average value of the past two years (3.35 GW/y) [37]. For a more comprehensive analysis, the demand for greening of grey H_2 production sites and power sector (e.g., transportation, heating) should be, hence, taken into account.

According to Armaroli et al. [44], they would require an additional 69.19 TWh, making the transition to green H_2 more challenging due to the high burden on the renewable grid. In fact, although the use of green H_2 would provide the greatest environmental benefits, the greening of the steel sector, grey H_2 production sites and power sector would require a total of 104.98 TWh of PV renewable energy, which translates to a total of 946 km^2 of surface covered by PV panels alone, 2.5 times the surface of Garda Lake. Similar difficulties were found in the analysis of the Swedish steel sector conducted by Toktarova et al. [31] in which the implementation of an H_2-DR/EAF plant to replace the national BF/BOF plants would lead to a near-zero emissions scenario at the expense of a significant renewable energy demand. Specifically, in the case where total Swedish steel production was maintained at 3 Mton of steel produced per year with an excess of 6 Mton of H_2-HBI for export, the additional energy demand of the steel sector would account for 33 TWh/year.

Hence, as also suggested by Toktarova et al. [31], pathway 2 appears to be a more reasonable scenario, being that the bio-CH_4-DR plants are able to emit, at full capacity, less than the current value of the BF/BOF plant (3.6 $Mton_{CO2}$ at 6 $Mton_{DRI\text{-}IT}$ vs. 7.74 $Mton_{CO2}$ at 3 $Mton_{steel}$) and require 17 times less energy demand than H_2-DR plants, also avoiding NG import as assumed in pathway 0. On the other hand, the application of biomethane could be limited by competition with other sectors that will consider it a viable gas stream for emissions mitigation [39].

Finally, regardless of the path, the Italian steel sector is already in a favorable situation compared to other major steel-producing countries, which could even be strengthened by the transformation of the BF/BOF plant into a conventional electric steel mill rather than the subsequent creation of the DR/EAF grid because of the little benefit brought by the latter compared to the challenge that is associated to each path. Indeed, to reflect the current decision-making process of the German and Japanese steel industry, Arens et al. [29] and Kuramochi et al. [44] analyzed the transition from an integrated plant to an innovative steel process (e.g., HIsarna, ULCORED, Top Gas Recycle Blast Furnace, CCS) without considering a significant increase in scrap route share. Their studies showed that the mitigation of emissions from the steel industries in the respective countries is highly limited by 2030 and would not meet the European and German climate targets by 2030 due to the high intensity of steel production from BF/BOF, even when applying the best available technologies.

5. Conclusions

This study explored the benefits and limitations of converting the Italian steel sector to a DR/EAF production following the closure of the Taranto BF/BOF plant in 2030 and its conversion to a DR plant with full capacity reached by 2050. Imported NG, green H_2 and biomethane were chosen to be studied as possible gas streams to be fed into the DR plant as reducing agents and into the EAF plants as methane substitutes. Finally, the metallic charge of the EAFs was assumed to be covered for 10 wt.% by DRI.

The results showed that, regardless of the chosen pathway, the closure of the integrated steel mill in 2030 would result in not only the highest emission reduction from the current value (71%) but also the lowest specific emissions, amounting to 1/10 of the IEA's SDS target. Similarly, the specific emission of each pathway in 2050 would remain far from the respective year's climate target, even considering a constant steel production of 22 Mton and a DRI production of 6 Mton, at the expense, however, of doubling the amount of imported iron ore from the 2022 value and increasing energy demand due to DR plant operation and reduced gas production.

The exploitation of imported NG and biomethane as a reducing gas in DR plants provided similar results in 2050 with a relative increase in emissions of 158% and 125% over 2030 due to plant operation and domestic biomethane production. In contrast, green H_2 would be able to further reduce emissions by 32% but at the cost of increasing the sector energy demand by 216% due to hydrogen production over the same time period.

Finally, the sensitivity analysis showed that, although in principle each of the pathways could be feasible, the creation and increase in the production share of the DR/EAF grid from 2030 onward are vitally linked to the establishment of a rigorous national renewable energy policy and the creation of biomethane production capacity as well as meeting the ambitious 2050 targets for the Italian renewable energy grid.

Author Contributions: Conceptualization, G.D., D.M. and C.M.; methodology, G.D.; software, G.D.; validation, G.D., D.M. and C.M.; formal analysis, G.D., D.M. and C.M.; investigation, G.D.; resources, G.D.; data curation, G.D.; writing—original draft preparation, G.D.; writing—review and editing, G.D., D.M. and C.M.; visualization, G.D.; supervision, D.M. and C.M. All authors have read and agreed to the published version of the manuscript.

Funding: This research received no external funding.

Data Availability Statement: The data presented in this study are available on request from the corresponding author (accurately indicate status).

Conflicts of Interest: The authors declare no conflicts of interest.

Nomenclature

List of acronyms	
BF	Blast furnace
Bio-CH_4	Biomethane
BOF	Basic oxygen furnace
CCS	Carbon capture and storage
DR	Direct reduction
DRI	Direct reduced iron
DRI-IT	Domestically produced DRI
DRI-IT/EAF	Electric steel plants exploiting (charging or fed by) domestic DRI
EAF	Electric arc furnace
HBI	Hot briquetted iron
NG	Natural gas
PV	Photovoltaic
SDS	Sustainable development scenario
List of Symbols	
i	Pathway investigated (0: natural gas; 1: green hydrogen; 2: biomethane)
j	Specific production cycle (integrated cycle, scrap recycling and DR)
P	Amount of steel produced per year [Mton]
s	Specific CO_2 emission [ton_{CO2}/ton_{steel}]
t	Year
TE	Total CO_2 emission [Mton]
TEC	Total electrical consumption [TWh]
ξ	Amount of production capacity increase of the DR plant per year
χ	Share of the specific production cycle

References

1. World Steel Association. *World Steel in Figures 2023*; World Steel Association: Bruxelles, Belgium, 2023.
2. International Energy Agency. *Emissions Measurement and Data Collection for a Net Zero Steel Industry*; International Energy Agency: Paris, France, 2023.
3. United Nations. Adoption of the Paris Agreement. In Proceedings of the Conference of the Parties, Twenty-First Session, Paris, France, 30 November–11 December 2015; pp. 1–32.
4. International Energy Agency. *Iron and Steel Technology Roadmap towards More Sustainable Steelmaking Part of the Energy Technology Perspectives Series*; International Energy Agency: Paris, France, 2020.
5. Kieush, L.; Rieger, J.; Schenk, J.; Brondi, C.; Rovelli, D.; Echterhof, T.; Cirilli, F.; Thaler, C.; Jaeger, N.; Snaet, D.; et al. A Comprehensive Review of Secondary Carbon Bio-Carriers for Application in Metallurgical Processes: Utilization of Torrefied Biomass in Steel Production. *Metals* **2022**, *12*, 2005. [CrossRef]
6. Chen, Y.; Zuo, H. Review of Hydrogen-Rich Ironmaking Technology in Blast Furnace. *Ironmak. Steelmak.* **2021**, *48*, 749–768. [CrossRef]
7. Perpiñán, J.; Peña, B.; Bailera, M.; Eveloy, V.; Kannan, P.; Raj, A.; Lisbona, P.; Romeo, L.M. Integration of Carbon Capture Technologies in Blast Furnace Based Steel Making: A Comprehensive and Systematic Review. *Fuel* **2023**, *336*, 127074. [CrossRef]
8. Bhaskar, A.; Assadi, M.; Nikpey Somehsaraei, H. Decarbonization of the Iron and Steel Industry with Direct Reduction of Iron Ore with Green Hydrogen. *Energies* **2020**, *13*, 758. [CrossRef]
9. Fischedick, M.; Marzinkowski, J.; Winzer, P.; Weigel, M. Techno-Economic Evaluation of Innovative Steel Production Technologies. *J. Clean. Prod.* **2014**, *84*, 563–580. [CrossRef]
10. Lopez, G.; Farfan, J.; Breyer, C. Trends in the Global Steel Industry: Evolutionary Projections and Defossilisation Pathways through Power-to-Steel. *J. Clean. Prod.* **2022**, *375*, 134182. [CrossRef]
11. Zhang, X.; Jiao, K.; Zhang, J.; Guo, Z. A Review on Low Carbon Emissions Projects of Steel Industry in the World. *J. Clean. Prod.* **2021**, *306*, 127259. [CrossRef]
12. Net Zero by 2050—Analysis—IEA. Available online: https://www.iea.org/reports/net-zero-by-2050 (accessed on 8 December 2023).
13. Mapelli, C.; Dall'Osto, G.; Mombelli, D.; Barella, S.; Gruttadauria, A. Future Scenarios for Reducing Emissions and Consumption in the Italian Steelmaking Industry. *Steel Res. Int.* **2022**, *93*, 2100631. [CrossRef]
14. DiGiovanni, C.; Li, D.; Ng, K.W.; Huang, X. Ranking of Injection Biochar for Slag Foaming Applications in Steelmaking. *Metals* **2023**, *13*, 1003. [CrossRef]
15. Echterhof, T.; Pfeifer, H. Study on Biochar Usage in the Electric Arc Furnace. In Proceedings of the 2nd International Conference Clean Technologies in the Steel Industry, Budapest, Hungary, 26–28 September 2011.
16. Sustainable EAF Steel Production (GREENEAF)—Publications Office of the EU. Available online: https://op.europa.eu/en/publication-detail/-/publication/e7dc500c-82de-4c2d-8558-5e24a2d335fb/language-en (accessed on 6 July 2023).

17. Robinson, R.; Brabie, L.; Pettersson, M.; Amovic, M.; Ljunggren, R. An Empirical Comparative Study of Renewable Biochar and Fossil Carbon as Carburizer in Steelmaking. *ISIJ Int.* **2022**, *62*, 2522–2528. [CrossRef]
18. Meier, T.; Hay, T.; Echterhof, T.; Pfeifer, H.; Rekersdrees, T.; Schlinge, L.; Elsabagh, S.; Schliephake, H. Process Modeling and Simulation of Biochar Usage in an Electric Arc Furnace as a Substitute for Fossil Coal. *Steel Res. Int.* **2017**, *88*, 1600458. [CrossRef]
19. Cardarelli, A.; De Santis, M.; Cirilli, F.; Barbanera, M. Computational Fluid Dynamics Analysis of Biochar Combustion in a Simulated Ironmaking Electric Arc Furnace. *Fuel* **2022**, *328*, 125267. [CrossRef]
20. Biochar for a Sustainable EAF Steel Production (GREENEAF2)—Publications Office of the EU. Available online: https://op.europa.eu/en/publication-detail/-/publication/7198c147-22b2-11e9-8d04-01aa75ed71a1/language-en (accessed on 6 July 2023).
21. Midrex Technologies Inc. *2022 World Direct Reduction Statistics*; Midrex Technologies Inc.: Charlotte, NC, USA, 2023.
22. GMK Center. Global DRI Production Will Increase to 175 Million Tons by 2030—WSD. Available online: https://gmk.center/en/news/global-dri-production-will-increase-to-175-million-tons-by-2030-wsd/ (accessed on 12 December 2023).
23. Zaini, I.N.; Nurdiawati, A.; Gustavsson, J.; Wei, W.; Thunman, H.; Gyllenram, R.; Samuelsson, P.; Yang, W. Decarbonising the Iron and Steel Industries: Production of Carbon-Negative Direct Reduced Iron by Using Biosyngas. *Energy Convers. Manag.* **2023**, *281*, 116806. [CrossRef]
24. Federacciai. *La Siderugia Italiana in Cifre 2022*; Federacciai: Milano, Italy, 2022.
25. GMK Center DRI d'Italia Will Contribute to the Decarbonization of the Country's Steel Sector. Available online: https://gmk.center/en/news/dri-ditalia-will-contribute-to-the-decarbonization-of-the-countrys-steel-sector/ (accessed on 12 December 2023).
26. GMK Center DRI d'Italia Will Build a DRI Producing Plant in Taranto. Available online: https://gmk.center/en/news/dri-ditalia-will-build-a-dri-producing-plant-in-taranto/ (accessed on 12 December 2023).
27. Mandova, H.; Patrizio, P.; Leduc, S.; Kjärstad, J.; Wang, C.; Wetterlund, E.; Kraxner, F.; Gale, W. Achieving Carbon-Neutral Iron and Steelmaking in Europe through the Deployment of Bioenergy with Carbon Capture and Storage. *J. Clean. Prod.* **2019**, *218*, 118–129. [CrossRef]
28. Arens, M.; Worrell, E.; Eichhammer, W.; Hasanbeigi, A.; Zhang, Q. Pathways to a Low-Carbon Iron and Steel Industry in the Medium-Term—The Case of Germany. *J. Clean. Prod.* **2017**, *163*, 84–98. [CrossRef]
29. Bhaskar, A.; Abhishek, R.; Assadi, M.; Somehesaraei, H.N. Decarbonizing Primary Steel Production: Techno-Economic Assessment of a Hydrogen Based Green Steel Production Plant in Norway. *J. Clean. Prod.* **2022**, *350*, 131339. [CrossRef]
30. Toktarova, A.; Karlsson, I.; Rootzén, J.; Göransson, L.; Odenberger, M.; Johnsson, F. Pathways for Low-Carbon Transition of the Steel Industry—A Swedish Case Study. *Energies* **2020**, *13*, 3840. [CrossRef]
31. Hong, W.Y. A Techno-Economic Review on Carbon Capture, Utilisation and Storage Systems for Achieving a Net-Zero CO_2 Emissions Future. *Carbon Capture Sci. Technol.* **2022**, *3*, 100044. [CrossRef]
32. Renzulli, P.; Notarnicola, B.; Tassielli, G.; Arcese, G.; Di Capua, R. Life Cycle Assessment of Steel Produced in an Italian Integrated Steel Mill. *Sustainability* **2016**, *8*, 719. [CrossRef]
33. Madias, J. Electric Furnace Steelmaking. In *Treatise on Process Metallurgy*; Elsevier: Amsterdam, The Netherlands, 2014; Volume 3, pp. 271–300.
34. Gaeta, M.; Nsangwe Businge, C.; Gelmini, A. Achieving Net Zero Emissions in Italy by 2050: Challenges and Opportunities. *Energies* **2021**, *15*, 46. [CrossRef]
35. Bianco, V.; Cascetta, F.; Nardini, S. Analysis of Technology Diffusion Policies for Renewable Energy. The Case of the Italian Solar Photovoltaic Sector. *Sustain. Energy Technol. Assess.* **2021**, *46*, 101250. [CrossRef]
36. TERNA. *Rapporto Mensile Sul Sistema Elettrico—November 2023*; TERNA: Rome, Italy, 2023.
37. Noussan, M.; Negro, V.; Prussi, M.; Chiaramonti, D. The Potential Role of Biomethane for the Decarbonization of Transport: An Analysis of 2030 Scenarios in Italy. *Appl. Energy* **2024**, *355*, 122322. [CrossRef]
38. European Biogas Association. *Biomethane Production Potentials in the EU*; European Biogas Association: Brussels, Belgium, 2022.
39. Ghosh, A.M.; Vasudevan, N.; Kumar, S. *Compendium: Energy-Efficient Technology Options for Direct Reduction of Iron Process (Sponge Iron Plants)*; Taylor & Francis: New Delhi, India, 2021.
40. Siderweb Preridotto, Progetti Nazionali in Stallo. Available online: https://www.siderweb.com/articoli/top/719730-preridotto-progetti-nazionali-in-stallo (accessed on 27 December 2023).
41. GMK Center Ex-Ilva: Steel Plant in Taranto Continues the Fight for the Future. Available online: https://gmk.center/en/posts/ex-ilva-steel-plant-in-taranto-continues-the-fight-for-the-future/ (accessed on 27 December 2023).
42. Pauliuk, S.; Milford, R.L.; Müller, D.B.; Allwood, J.M. The Steel Scrap Age. *Environ. Sci. Technol.* **2013**, *47*, 3448–3454. [CrossRef] [PubMed]

43. Armaroli, N.; Barbieri, A. The Hydrogen Dilemma in Italy's Energy Transition. *Nat. Italy* **2021**. [CrossRef]
44. Kuramochi, T. Assessment of Midterm CO_2 Emissions Reduction Potential in the Iron and Steel Industry: A Case of Japan. *J. Clean. Prod.* **2016**, *132*, 81–97. [CrossRef]

metals

MDPI

Article

Impact of Injection Rate on Flow Mixing during the Refining Stage in an Electric Arc Furnace

Orlando Ugarte [1,*], Neel Busa [1], Bikram Konar [2], Tyamo Okosun [1] and Chenn Q. Zhou [1]

1 Center for Innovation through Visualization and Simulation (CIVS), Steel Manufacturing Simulation and Visualization Consortium (SMSVC), Purdue University Northwest, Hammond, IN 46323, USA; busa@pnw.edu (N.B.); tokosun@pnw.edu (T.O.); czhou@pnw.edu (C.Q.Z.)

2 EVRAZ North America, Regina, SK S4P 3C7, Canada; bikram.konar@evrazna.com

* Correspondence: ougarte@pnw.edu; Tel.: +1-219-989-2089

Abstract: During the refining stage of electric arc furnace (EAF) operation, molten steel is stirred to facilitate gas/steel/slag reactions and the removal of impurities, which determines the quality of the steel. The stirring process can be driven by the injection of oxygen, which is carried out by burners operating in lance mode. In this study, a computational fluid dynamics (CFD) platform is used to simulate the liquid steel flow dynamics in an industrial-scale scrap-based EAF. The CFD platform simulates the three-dimensional, transient, non-reacting flow of the liquid steel bath stirred by oxygen injection to analyze the mixing process. In particular, the CFD study simulates liquid steel flow in an industrial-scale EAF with three asymmetric coherent jets, which impacts the liquid steel mixing under different injection conditions. The liquid steel mixing is quantified by defining two variables: the mixing time and the standard deviation of the flow velocity. The results indicate that the mixing rate of the bath is determined by flow dynamics near the injection cavities and that the formation of very low-velocity regions or 'dead zones' at the center of the furnace and the balcony regions prevents flow mixing. This study includes a baseline case, where oxygen is injected at 1000 SCFM in all the burners. Two sets of cases are also included: The first set considers cases where oxygen is injected at a reduced and at an increased uniform flow rate, 750 and 1250 SCFM, respectively. The second set considers cases with non-uniform injection rates in each burner, which keep the same total flow rate of the baseline case, 3000 SCFM. Comparison between the two sets of simulations against the baseline case shows that by increasing the uniform flow rate from 1000 to 1250 SCFM, the mixing time is reduced by 10.9%. Moreover, all the non-uniform injection cases reduce the mixing time obtained in the baseline case. However, the reduction in mixing times in these cases is accompanied by an increase in the standard deviations of the flow field. Among the non-uniform injection cases, the largest reduction in mixing time compared to the baseline case is 10.2%, which is obtained when the largest flow rates are assigned to coherent jets located opposite each other across the furnace.

Keywords: CFD; EAF; mixing rate; liquid bath; molten steel; refining; injection rate

1. Introduction

An electric arc furnace (EAF) is a low-carbon emission steelmaking route widely adopted by the steel industry worldwide. Specifically, the EAF route can reduce the carbon emissions produced by the traditional blast furnace/basic oxygen furnace (BF-BOF) route by up to 63% [1]. Since the iron and steel industry produces ~6% of the global CO_2 emissions, and 71% of the iron and steel production utilizes the BF-BOF route [1,2], transition to EAF as the main route in the steelmaking industry is necessary in order to reach the carbon emission goals established for 2050. The flexibility of EAF operation, along with the control over steel temperature and grades it provides, enables the production of multiple steel products. The EAF primarily uses electrical energy with some chemical energy to melt and refine ferrous scrap and produce steel.

A typical EAF operation can be distinguished into four stages: charging, preheating, melting of scrap/iron substrates, and refining and tapping of liquid steel. In the preheating stage, the temperature of the ferrous charge is increased to facilitate its melting. In the melting stage, the ferrous charge is exposed to chemical reactions (due to oxidation reactions and burners) and electrical heating that melt the charge and produce a molten bath. In modern high-power EAFs, electrical heating accounts for 50–60% of the energy supplied. The refining stage refers to the period during which the molten bath is exposed to chemical–physical processes, such as supersonic oxygen injection into the steel, to refine and obtain (tapping) the desired liquid steel chemistry and temperature.

The refining stage reactions are controlled by the local concentration of species and flow properties of the liquid steel. These reactions can be enhanced by stirring forces as they increase the flow by mixing. For this reason, technologies such as O_2 injection through lances, bottom gas injection (BGI), and electromagnetic stirring (EMS) are applied in the refining stage [3]. The BGI technology is extensively applied in ladle furnaces. However, the application of BGI to EAF operation has challenges. These challenges are related to the height of the molten steel, which is variable in the EAF case, and the low ratio of liquid height to furnace diameter seen in EAF as compared to the ladle (3–5 times lower). As a result, large spouts may form at the bath surface in EAF, which affects the arc stability [3]. The EMS concept is an old one, but it has not been extensively commercialized in EAFs. Electric arcs in EAF generate electromagnetic forces, although these forces are weak. The installation of coils at the bottom of the EAF, where low-frequency current is applied, has demonstrated a significant stirring impact on the molten bath [4]. Recent implementations of EMS in EAF have shown significant improvements in thermal stratification, decarburization rates, and electrode consumption [5]. However, as mentioned earlier, EMS technology is not widely implemented. A well-established technology used in the refining stage of EAF is oxygen injection from lances. In this case, oxygen is injected into the molten steel through jets located above the liquid steel in order to stir it, remove impurities, and provide the right slag foaming. Once the injected oxygen dissolves in the liquid steel, the oxygen reacts with the carbon in the liquid to generate in-bath oxygen and CO bubbles, which generate turbulence and further stir the molten bath [6].

All these technologies aim to increase mixing in the molten bath to promote reactions during the refining stage. Although it is known that mixing is controlled by forces such as electromagnetics forces, buoyancy, momentum provided by flow injection (BGI or top oxygen injection), and CO bubble dynamics, the impact of specific forces on bath mixing is not well understood. Research on mixing efficiency in multiphase fluid flows has been conducted in numerous scenarios, such as channels and serpentine mixers [7–10], as well as in steelmaking processes. For instance, Li et al. [11] utilized the volume of fluid (VOF) multiphase model integrated with a discrete phase model (DPM) to describe the gas and liquid two-phase flow in a steelmaking converter that included top and bottom blowing. The study concluded that the buoyancy-driven bubbles are the driving force behind the majority of the evolved stirring energy in the converter. Interestingly, it was found that mixing efficiency was higher when three bottom tuyeres were used instead of two or four. This application is similar to the phenomenon produced in the refining stage in the EAF, where continuous blowing of oxygen into the liquid bath forms bubbles and induces stirring. Duan and Wei [12] analyzed the flow in an argon oxygen decarburization (AOD) converter with combined side–top blowing. The fluid mixing characteristics were investigated by studying the effect of rotation of the side-blowing gas jet and the influence of varying the blowing volume. The results indicate that improved mixing efficiency can be achieved by rotating the gas jet under the same blowing volume, which also improves the agitation.

Extensive flow analysis has been conducted in ladles, where water models have provided data needed for phenomena understanding and model validation. It is possible to translate this knowledge to the EAF operation, as similar flow features can be observed in both ladles and EAF (i.e., [13]). Zhu et al. [14] used an experimental water model and mathematical modeling to study mixing phenomena in a gas-stirred ladle. Similar to Li et al. [11], Zhu et al. [14] showed that the arrangement of tuyeres played a major role in the flow mixing. In particular, it was found that a single tuyere located at an off-centric position led to the shortest mixing time. The results also showed a significant impact of the angle of blowing on bath mixing and that mixing is enhanced by increased gas flow rate, although the latter had a secondary role. The study proposed a correlation to determine the mixing time in the ladle configuration. Amaro-Villeda et al. [15] also analyzed the flow mixing in ladles and added the effect of slag properties into the analysis. The results showed that slag thickness has a negative impact on mixing, as it leads to higher energy dissipation. Also, it was found that the range of energy dissipated to the top layer of the ladle is 4–12%. The authors concluded that due to the multiple injection parameters, such as gas flow rate, number of nozzles, and position of nozzles, it is difficult to establish an optimal number of injectors for flow mixing. Cheng et al. [16] investigated the mixing process in a ladle that was scaled down to 1:3 from a 150 t industrial ladle. This configuration considered both side and bottom blowing. In this study, velocity fields and diffusion paths obtained by computational fluid dynamics (CFD) simulations were compared with experimental data, showing good agreement. The results also showed that mixing time depends on agitation power and the associated uniformity of the flow velocity. Overall, it was concluded that shorter mixing times were obtained with a side-blowing configuration compared with bottom blowing.

Chen et al. [6] developed a CFD platform to compute the refining process in industrial-scale EAFs. The model computed the coherent jets and liquid baths separately, avoiding the restrictions set by the supersonic flow on the computational time needed during the refining simulations. The models were then integrated back to compute the in-bath decarburization process. The results obtained with this methodology showed that bubble stirring is a major mechanism for promoting flow homogenization in the molten steel bath. Also, a significant impact of oxidation reactions on the liquid temperature was identified. Decarburization reactions were observed mostly around the cavities produced by the penetrating jets. Li [17] analyzed the mixing process in an electric arc furnace with bottom stirring. This study considered both computational and experimental methods to analyze a 1:7 model of a 30-ton EAF. The impact of parameters such as diameter and location of plugs on flow mixing was analyzed. The results showed that off-center blowing of gas into the liquid bath improved the intensity of stirring and the mixing rate. Moreover, the mixing rate did not show a significant increase with plug diameter, but increasing the number of plugs had a positive impact on the flow homogenization.

The research mentioned above quantifies the metrics for mixing and homogenization in a liquid bath, but it also demonstrates that the mixing characteristics are somehow specific to the particular scenario. In order to determine the conditions for optimal flow mixing in EAF, it is necessary to model the EAF with the conditions and geometry to be used in the actual operation. This study aims to determine the flow features and the conditions for optimal mixing in an industrial-scale EAF, where oxygen is injected through coherent jets. Since the flow rate can be varied in the co-jets within a certain range, this study analyzes the impact of having different flow rates in the co-jets during EAF operation. Different scenarios are computed to understand how the injection rates modify the flow field and, as a result, the mixing process. A CFD platform introduced by Chen et al. [6] is used to perform the calculations. This investigation focuses on the flow field developed in the steel bath as a result of the oxygen injection, and thus the reactions occurring during the refining stage are not considered ('cold flow' approach). The simulations do consider the combustion reactions of the shrouded flame in the coherent jets.

2. Methodology

2.1. Description of the Integrated CFD Platform for EAF Refining Simulations

This section contains a brief description of the computational model used to simulate the oxygen injection in liquid bath scenarios. For more information, including details of the model validation and CFD implementation, the reader is referred to [6].

The model uses three steps to set up the physical boundaries of the domain and perform the computations. These steps are defined according to three physical processes occurring concurrently during the EAF refining: (1) oxygen blowing into the liquid bath via co-jet burners, (2) the interaction between the coherent jet and the liquid steel, and (3) the stirring of the bath due to high-intensity jet injection. This methodology ensures the simulation accuracy of the refining stage without compromising the computation resources, as the supersonic regime is restricted to the first step only. These three steps are summarized as follows:

1. The supersonic coherent jet is simulated first based on the injection conditions of the burners operating in lance mode. This simulation assumes a steady-state condition, as the impact of the liquid bath on co-jet operation is expected to be minor. This step provides velocity profiles and composition of the injecting flow from the tip of the burner to the surface of the bath.
2. Outputs from the coherent jet simulation are used to estimate the cavities formed by the jets on the surface of the liquid bath.
3. The computational domain for the refining simulation is created based on the actual geometry of the industrial-scale EAF and the geometry of cavities calculated in step 2. This domain includes the liquid bath only. A transient simulation is performed in the computational domain where oxygen is injected at the cavities, at the rate provided by the coherent jet simulation solution.

The refining CFD platform used in step 3 is able to compute multiple species and reactions. However, in this study, the refining reactions are not computed, and it is assumed that only one liquid species is present in the liquid bath (steel). Next, each of the steps listed above is further described.

2.2. Simulation of the Supersonic Coherent Jet Burners (Steady State Simulation)

The supersonic coherent jet simulation considers an open space under furnace conditions. The coherent jet injection is solved as a compressible, non-isothermal, steady-state flow. The solver used is a compressible solver implemented in ANSYS Fluent version 2019. The continuity equation solved for the coherent jet is as follows:

$$\nabla.\left(\rho\vec{v}\right)=0 \tag{1}$$

The momentum equation is constructed as follows:

$$\nabla.\left(\rho\vec{v}\vec{v}\right)=-\nabla p+\nabla.\left(\overline{\overline{\tau}}\right)+\rho\vec{g}+\vec{F} \tag{2}$$

where ρ, $\vec{v}$, p, $\overline{\overline{\tau}}$, g, and $\vec{F}$ are the density, velocity vector, static pressure, stress tensor, gravity acceleration, and the external body force, respectively. Moreover, the energy conservation equation can be written as follows:

$$\nabla.\left[\vec{v}(\rho E+p)\right]=\nabla.\left[\left(k+\frac{c_p.\mu_t}{Pr_t}\right)\nabla T-\sum_j h_j\vec{J_J}+\left(\overline{\overline{\tau}}_{eff}\,.\,\vec{v}\right)\right]+S_h, \tag{3}$$

Here, E, k, c_p, and μ_t are the total energy that corresponds to the sensible enthalpy h, thermal conductivity, specific heat, and turbulent viscosity, respectively. Pr_t is the turbulent Prandtl number (equal to 0.85) for the $k-\varepsilon$ turbulence model, $\vec{J_J}$ is the diffusion flux of substance j. S_h is the volumetric heat source, which includes the heat of chemical reactions. Note that even though the simulations do not include refining reactions, step 1 accounts for combustion reactions of the shrouded flame present in the co-jets. The turbulence viscosity is modeled as:

$$\mu_t = C_\mu \rho \frac{k^2}{\varepsilon}, \quad (4)$$

where the constant is modified in order to include the influence of entrained ambient gas [18] as follows:

$$C_\mu = \frac{0.09}{C_T}. \quad (5)$$

The modification of the turbulent viscosity is loaded into the CFD software via a user-defined function (UDF) code. Figure 1 shows the interaction of the coherent jet with the liquid bath. This interaction will generate a cavity in the surface of the liquid steel, which is computed in the second step of the CFD setup.

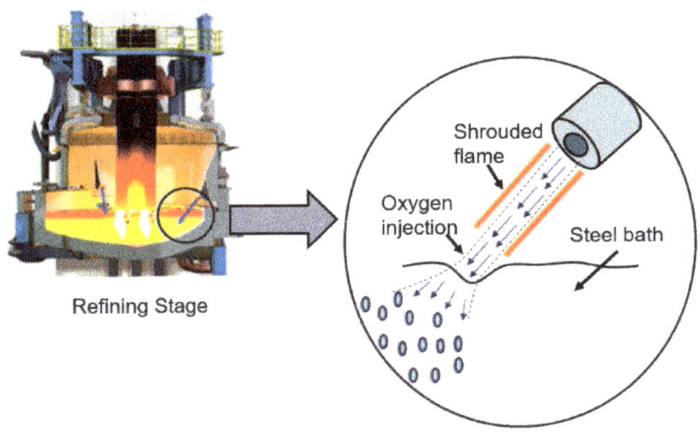

Figure 1. Cavity formation in the liquid bath due to jet impingement.

2.3. Calculation of the Cavity Produced by the Coherent Jet in the Liquid Bath

For the second step, a cavity is estimated on the surface of the liquid bath based on the impingement of the supersonic jet on the bath [19]. The jet momentum transfer and oxygen delivery are calculated in this step to estimate the cavity shape and size. Based on the energy balance calculation, the momentum transferred by the jet flow to the liquid steel is given by:

$$P_{s,avg} = \alpha \rho_{O_2} {v_{O_2}}^2 A = \frac{\alpha \rho_{O_2} A}{\rho_s} \left[\frac{1}{\Delta z} \int_{z2}^{z1} v_{O_2}(z) dz \right]^2 \quad (6)$$

The amount of deliverable oxygen from the jet to the liquid bath is expressed as:

$$m_{O_{2,avg}} = \frac{1}{\Delta z} \int_{z2}^{z1} m_{O_2}(z) dz \quad (7)$$

The cavity is used to inject oxygen into the liquid bath and induce in-bath stirring. About 6% of the momentum is transferred from the jet to the bath due to the dampening effects of the slag layer and viscosity effects [6]. The jet cavity design is integrated into the refining simulation, with the cavity being part of the boundaries of the EAF domain (to be shown in Section 2.3). The main characteristics needed from the coherent jet simulation are the velocity of the jet reaching the bath and the amount of oxygen reaching the bath. The shape of the cavity is assumed to be a 3D paraboloid which follows the mathematical expression

$$z = \frac{(x^2 + y^2)}{c}, \quad (8)$$

where c is the constant needed to be defined by a given volume and depth of the cavity. The volume of the cavity [20] can be obtained by

$$V = \frac{\pi \rho_j {v_j}^2 {d_j}^2}{4 g \rho_s}, \tag{9}$$

where ρ_j and ρ_s are the primary densities of the jet and liquid steel, and v_j and d_j are the jet velocity and jet diameter of the nozzle, respectively. The jet penetration depth is an empirical formula derived by Ishikawa et al. [21], which mathematically describes the indentation created by the coherent jet in the liquid bath. This is expressed as

$$D = \gamma_{h_0} e^{-\frac{\sigma_1 L}{\gamma_{h_0}{}^{cos\theta}}} \tag{10}$$

$$\gamma_{h_0} = \sigma_2 \left(\frac{\dot{V}}{nd\sqrt{3}} \right) \tag{11}$$

Here, L is the distance between the nozzle exit and bath, σ_1 and σ_2 are constants equal to 1.77 and 1.67, respectively, derived from an experiment analysis [22], θ is the angle of the jet inclination and n is the number of nozzles (1 in this study). The size and shape of the cavity vary according to the velocity of the jet reaching the bath, the diameter of the burner nozzle, and the density of the liquid bath.

2.4. Stirring of the Bath Due to Jet Injection (Transient Simulation)

Finally, the third part of the CFD platform is the simulation of the region containing the liquid steel bath generated in the furnace after all the scrap melts. The simulation incorporates the solution of the Navier–Stokes equations into a finite volume scheme by using the ANSYS Fluent platform. Specifically, the computational solution is based on a Eulerian multi-phase, incompressible approach, where the primary phase is the molten liquid, and the secondary phase is the injected oxygen. The simulation uses the standard *k-e* model with standard wall functions. The boundary conditions for oxygen injection are provided by the coherent jet solution, and these are included in the refining simulation with a user-defined function. Figure 2 shows the computational domain considered during the refining simulations on the right side of the figure. This domain includes the cavities produced by the coherent jets.

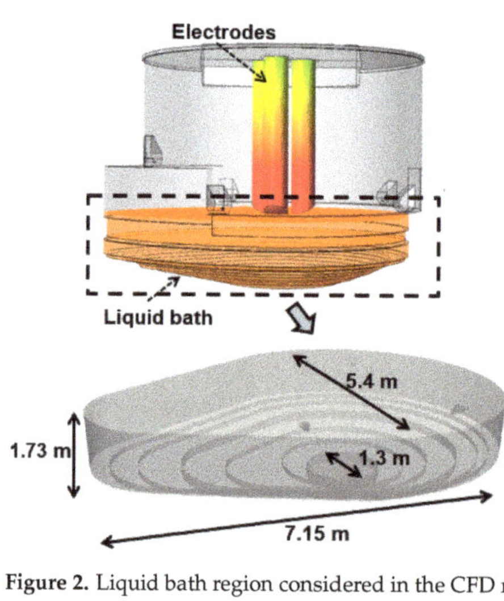

Figure 2. Liquid bath region considered in the CFD refining simulations, including the dimensions of the CFD domain.

2.5. Grid Sensitivity Study

A grid resolution study was performed on the computational setup shown in Figure 2. Namely, a case with the same operation conditions was repeated by using a coarse (0.3 million cells), a base (0.6 million cells), and a fine grid (1.3 million cells). The results obtained at a plane located 0.4 m from the top (close to the oxygen injection burners) are shown in Figure 3. The top set of contours shows the liquid velocity magnitude, whereas the lower ones show the velocity in the vertical direction. All three grids show similar flow structures. In particular, regions with large velocity magnitudes are observed between burners 1 and 3, and a reduced velocity flow develops near the balcony. Moreover, the three grids develop similar flow fields of the vertical velocity component (W_{Liquid}), which show the impact of the jets on the vertical acceleration.

Table 1. Results obtained with the three grid resolutions.

Grid	Number of Cells (Million)	Volume Averaged at Plane Located 0.4 m from the Surface (m^3)			CPU Hours
		U_{liquid}	V_{liquid}	W_{liquid}	
Coarse	0.3	−0.012	−0.020	0.0038	6464
Base	0.6	−0.009	−0.014	0.0027	6528
Fine	1.3	−0.006	−0.018	0.0022	7936

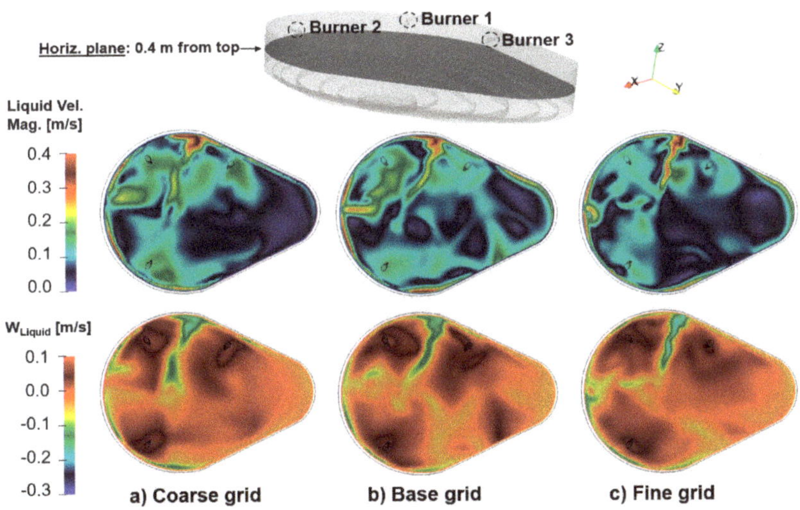

Figure 3. Liquid velocity magnitude and vertical velocity component computed on a plane located 0.4 m from the free surface (top boundary). Three grid resolutions are compared: (**a**) coarse grid, (**b**) base grid and (**c**) fine grid. Details of grid resolutions are shown in Table 1.

The main velocity components are listed in Table 1 for the three grid resolutions. Overall, we see that the order of the magnitude of the flow velocities is the same, with flow values mostly approaching the results of the fine grid. The results in Figure 3 and Table 1 indicate that the results are not further modified by grid refining: both the qualitative and quantitative CFD predictions do not change significantly when the grid is further refined, and the base grid is able to capture the flow features observed in the fine grid, where the number of cells is duplicated. The cases to be presented and discussed in this study were completed by using the base grid resolution.

2.6. Operation Conditions of Case Simulations

The simulations for this research use conditions of an actual EAF operation provided by EVRAZ North America. The EAF operation includes three co-jets working in lance mode. The shrouding gas in the co-jets is natural gas, which is injected into rings located around the central nozzle, where oxygen is injected. The gases are injected at room temperature. In the simulations of the liquid bath interacting with the injected oxygen, the gas corresponds to oxygen and the liquid to molten steel. The parameter values used in this study are listed in Table 2. The simulations include two main sets of results. The first set assumes the same injection rate in all three jets of the EAF. This includes the baseline case, where oxygen is injected at 1000 SCFM (standard cubic feet per minute), a case with reduced injection rates (750 SCFM), and a third case with increased injection rates (1250 SCFM). The second set of cases includes variable injection rates in the three jets, where the 750, 1000, and 1250 SCFM rates are assumed at each jet, in different order. The specific conditions of the variable injection rate cases will be discussed in Section 3.3.

Table 2. Parameters used in the simulations.

Name	Variables	Value
Jet cavities	Quantity	3
Oxygen injection	Flow rates	750, 1000, 1250 SCFM (0.47, 0.63, 0.78 kg/s)
	Mass fraction of oxygen	100%
Liquid Steel	Density	7500 kg/m^3
	Static temperature	1815 K (1542 C)
Coherent jet burner	Angle of inclination	45 degrees

3. Results

3.1. Impact of Flow Rate on the Coherent Jets' Penetration Depth

The coherent jet CFD model is used to obtain the flow velocity and oxygen mass fraction of the gas injected by the burners (operating in lance mode) when it reaches the molten bath. As explained in Section 2.1, the flow features produced by the injected gas are used to compute the cavities that jets produce in the bath, which are part of the computational domain used in the refining simulations. Figure 4 shows the velocity magnitude contours generated by the three injection rates considered in this study: 750, 1000, and 1250 SCFM. The dashed line shows the location at which the flow profile is obtained. The flow profile information at this location is then used in the calculation of the bath cavities by using the procedure explained in Section 2.3. The distance from the left boundary to the dashed line ($L = X/De = 30$, where De is the diameter of the jet) corresponds to the distance from the tip of the burner to the surface of the molten bath.

The results in Figure 4 show an increase in velocity magnitude as the flow rate increases, as expected. In all cases, the larger flow velocities are observed along the jet centerline. It is observed that the flow velocity is maintained until around $X/De = 50$ for all cases. This implies that flow entrainment generated as the jet interacts with the environment does not prevent the jets from transferring momentum to the liquid bath. Figure 5a compares the velocity profile at $X/De = 30$ for the three injection rates. Increasing the flow rate from 1000 to 1250 SCFM leads to a 16% increase in the maximum velocity, whereas reducing the flow rate from 1000 to 750 SCFM reduces the maximum velocity by 30%. Figure 5b shows the maximum and averaged axial velocities (z-component) for the three flow rates considered in this study. The average velocities approach a linear behavior, as they increase by 21.3 and 19.4% when increasing the O_2 rate from 750 to 1000 SCFM and from 1000 to 1250 SCFM, respectively.

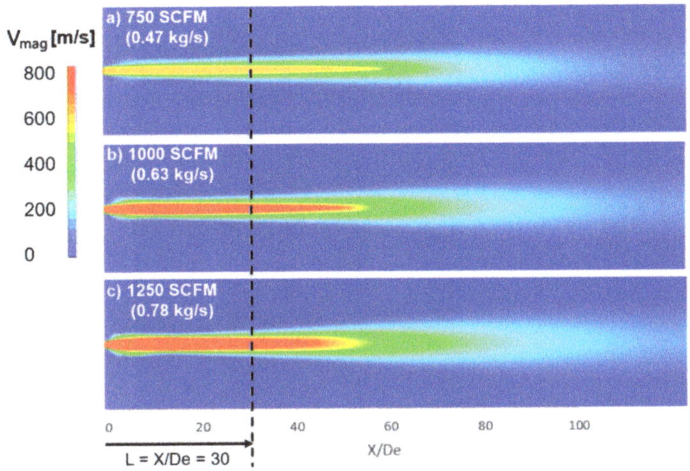

Figure 4. Injection of oxygen driven by coherent jets operating in lance mode for the three injection rates considered in this study.

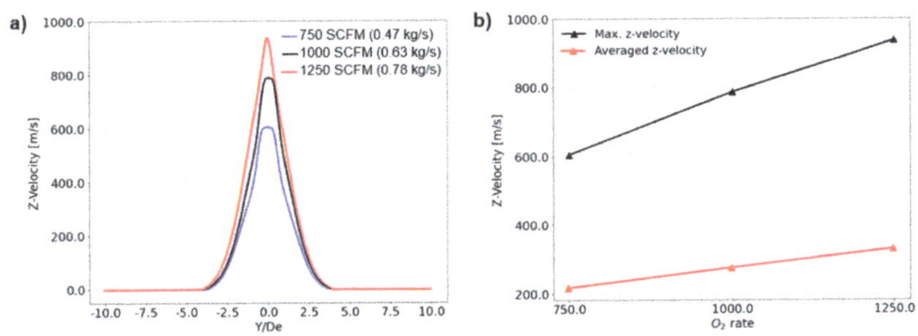

Figure 5. Results obtained by coherent jet model for O_2 rates considered in this study: (**a**) Z-velocity profiles computed vertically (Y/De direction) at X/De = 30, and (**b**) maximum and averaged z-velocity variation with O_2 injection.

The results obtained by the coherent jet CFD model, shown in Figures 4 and 5, are used to compute the cavities formed in the liquid bath. Figure 6 shows the cavities produced by the 750, 1000, and 1250 SCFM injection rates, and the variations in the volume and depth of the cavities with respect to the baseline case (1000 SCFM).

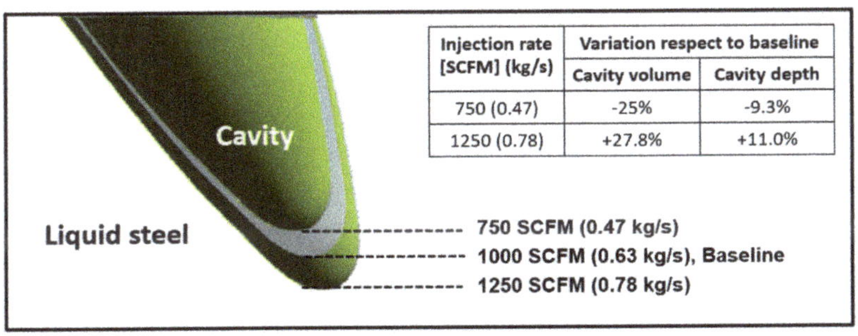

Injection rate [SCFM] (kg/s)	Variation respect to baseline	
	Cavity volume	Cavity depth
750 (0.47)	-25%	-9.3%
1250 (0.78)	+27.8%	+11.0%

Figure 6. The cavity formed due to jet impingement on a liquid bath for the three injection rates considered in this study.

3.2. *Baseline Results in the Steel Bath Domain*

The baseline case considers oxygen injection at a rate of 1000 SCFM in all the burners, which operate in lance mode. Figure 7 shows the location of the burners in Figure 7a and the flow patterns associated with such injection rates after 600 s in Figure 7b. The flow pattern shows large fluid velocities near the burners, dropping significantly as the flow changes direction due to interactions with the walls and the flow injected from adjacent burners. Overall, the flow velocity magnitude decreases from around 5 m/s to 0.01–0.5 m/s. Moreover, considering the entire domain, the average velocity is 0.024 m/s in the horizontal direction and 0.068 m/s in the vertical direction. These average velocities agree with those reported in Ref [3]. The flow pattern is particularly non-uniform. The main reasons for the non-uniformity of the flow are the geometry of the furnace, the asymmetric distribution of the injectors, and the angle at which the oxygen is injected.

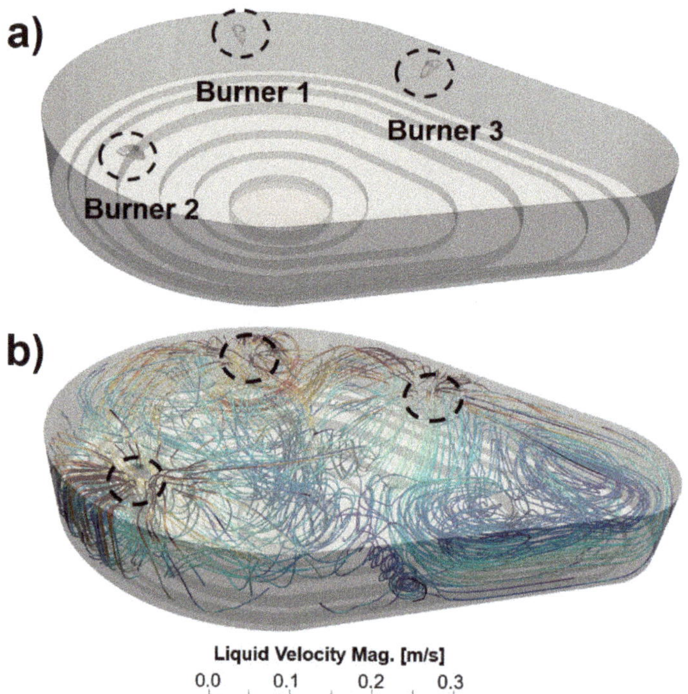

Figure 7. Flow characteristics in baseline case: (**a**) location of jet cavities produced by each of the burners, and (**b**) flow streamlines and velocity magnitude at 600 s.

Figure 8 shows the instantaneous velocity magnitude on a plane located 0.5 m from the top of the domain. The velocity contours are computed at t = 200 s, 400 s, and 600 s. The contours also include the velocity vectors of the flow field. Figure 8a through 8c show that the flow field does not change significantly along these time intervals. At all times, the flow velocities are larger near burner 1, as a result of the interaction of burners 1 and 3. By contrast, the flow injection near burner 2 decays quickly as this is the only injection point on the lower side of plane 1. Flow recirculation is shown at burner 1, and adjacent to burners 2 and 3. Also, the flow velocity near the balcony of the furnace (right end of plane 1) approaches zero, which can be attributed to the 'dead zone' in the liquid steel flow domain. Low-velocity regions or 'dead zones' are also formed on the lower wall of the plane (in the region opposite burner 3), near the center of the domain, and near burner 2. The percentages of the dead-zone volumes (defined as those regions where

V_{mag} < 0.05 m/s) are 6.3, 6.4, and 7.3% of the molten bath volume for time instants 200, 400, and 600 s, respectively. Overall, the flow features are consistent along the three time instances of the burner operation shown in Figure 8.

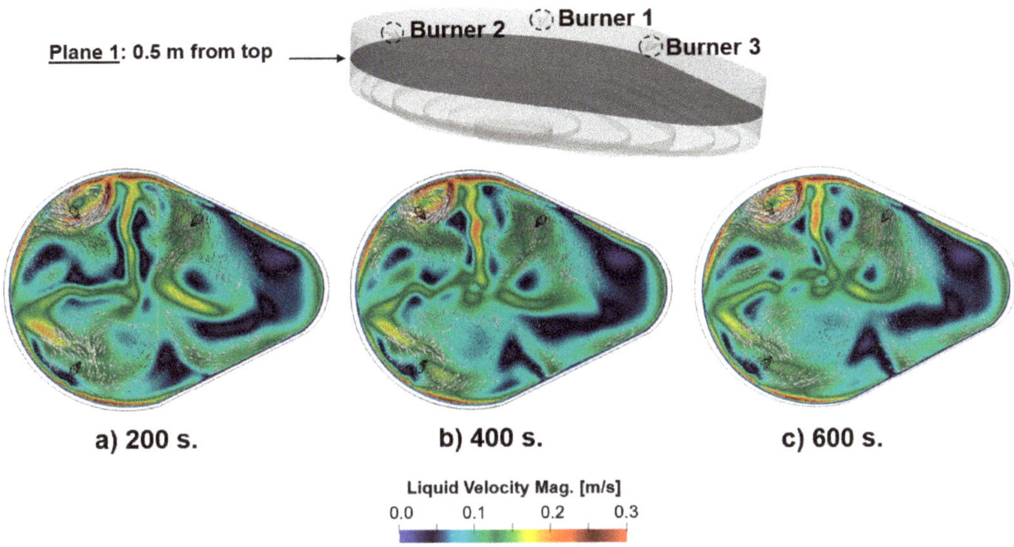

Figure 8. Time evolution of the velocity magnitude and flow vectors near the bath surface in the baseline case at (**a**) 200 s, (**b**) 400 s and (**c**) 600 s of operation.

Figure 9 shows the instantaneous velocity magnitude and flow vectors of the liquid steel on three planes, at t = 600 s. The locations of the planes are shown at the top of the figure. The results show that flow intensity is larger near the top planes of the domain (planes 1 and 2) where the momentum is transferred by the jets. The main features observed in plane 1 (Figure 9a) are also seen in plane 2 (Figure 9b). Namely, large velocities develop near burner 1 and on the upper side of planes 1 and 2, towards the left end. Plane 2 is further inside the furnace (Figure 9b) and, overall, it shows similar flow intensity as compared to plane 1. The flow velocities are reduced in plane 3, taken 1.2 m from the surface of the liquid bath. In plane 3, the velocity magnitude does not exceed 0.2–0.25 m/s, although recirculation patterns similar to those observed in planes 1 and 2 are maintained.

Figure 10 shows the mean velocity magnitude and the region of high-intensity flow in the planes indicated in Figure 9. Figure 10a shows that the mean velocity magnitude remains similar in the upper regions of the liquid bath, with a slight increase in the flow intensity when going from 0.5 to 0.85 m into the bath. Figure 10a also shows that the mean velocity magnitude decays by 28% on a plane located 1.2 m below the bath surface. This location is 0.15 m from the bottom of the furnace. Figure 10b, in turn, shows the percentage of the areas of planes 1–3 where the velocity magnitude is above 0.15 m/s. Here, it is shown that the region with significant flow intensity occupies ~12% of plane 2, and less than 8% of plane 3.

Figure 11 extends the analysis by considering two vertical planes, which are shown at the top of the figure. Figure 11a shows plane 4, which is oriented towards the balcony of the furnace. Plane 4 shows larger liquid velocities on the left side, near the locations of burners 1 and 2. The flow vectors on the left side of plane 4 show a vertical recirculation produced by the interaction of the flow injected at burners 1 and 2 with the bottom of the furnace. Therefore, the asymmetric distribution of the burners at the top of the liquid bath leads to the formation of recirculation structures in the horizontal and vertical directions near burners 1 and 2, as shown in planes 1–3 (Figure 9) and plane 4 (Figure 11a). Figure 11b shows larger flow velocities on plane 5 than on plane 4 due to the vicinity of burners 1 and 3, although the recirculation pattern in this plane is weaker than in plane 4. Planes 4 and 5 show large velocity gradients through the domain due to the asymmetric distribution of the burners and geometry of the EAF. Figure 12 confirms these observations by indicating increases in the mean velocity and size of the region with higher flow intensity in plane 5 with respect to plane 4 of 15 and 183%, respectively.

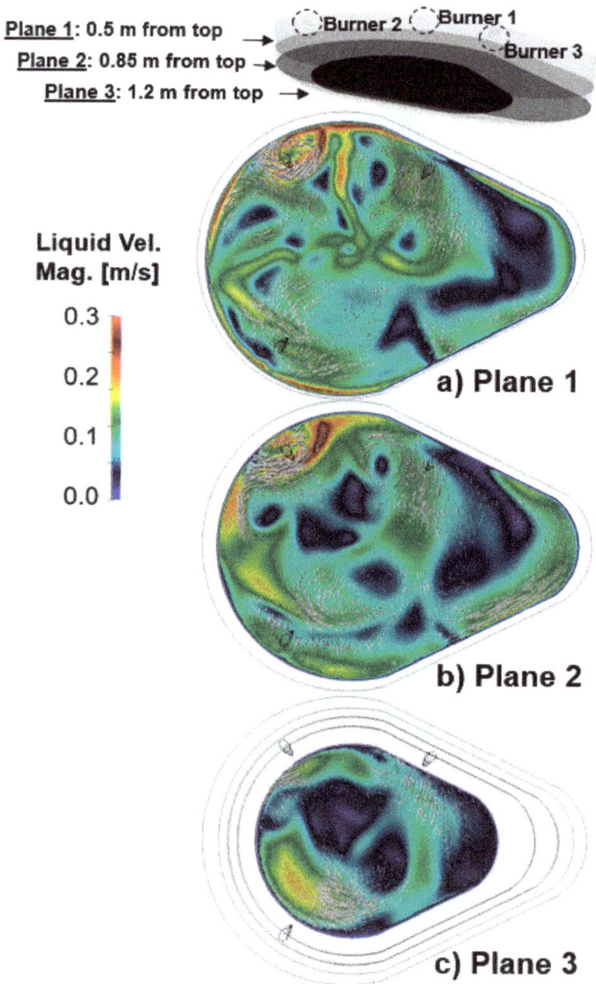

Figure 9. Velocity magnitude and flow vectors computed at (**a**) plane 1: 0.5 m from bath surface, (**b**) plane 2: 0.85 m from bath surface, and (**c**) plane 3: 1.2 m from bath surface, for baseline case.

The results in Figure 11 also suggest that the flow velocities and associated flow mixing would be increased by adding a fourth jet to the system. Namely, planes 4 and 5 show weak flows due to the presence of only one co-jet in the balcony. Since the balcony region is large, the stirring in the vertical direction is reduced away from burner 3, which is located near the wall of the balcony. Moreover, the balcony region extends the asymmetry of the bath horizontally, and the angle of injection of burner 3 (and burner 4 if it is included) could be modified so that the oxygen is injected towards the center of the balcony instead of the center of the liquid bath.

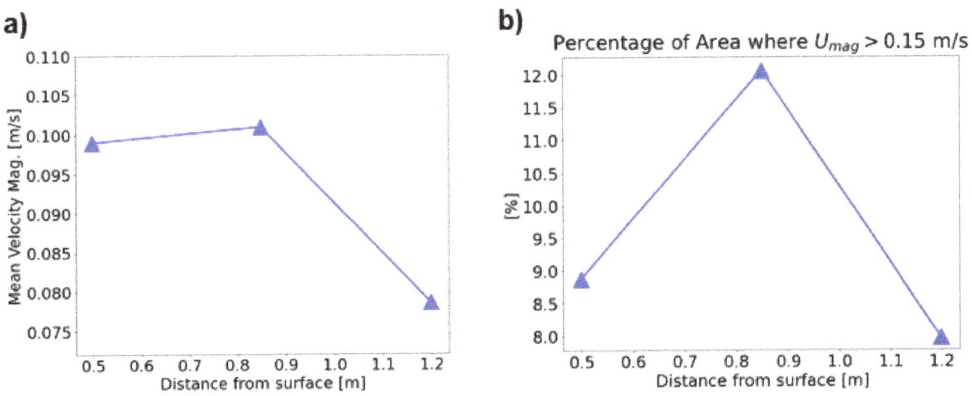

Figure 10. (**a**) Mean velocity magnitude and (**b**) percentage of area where the velocity magnitude is larger than 0.15 m/s, for the planes considered in Figure 9.

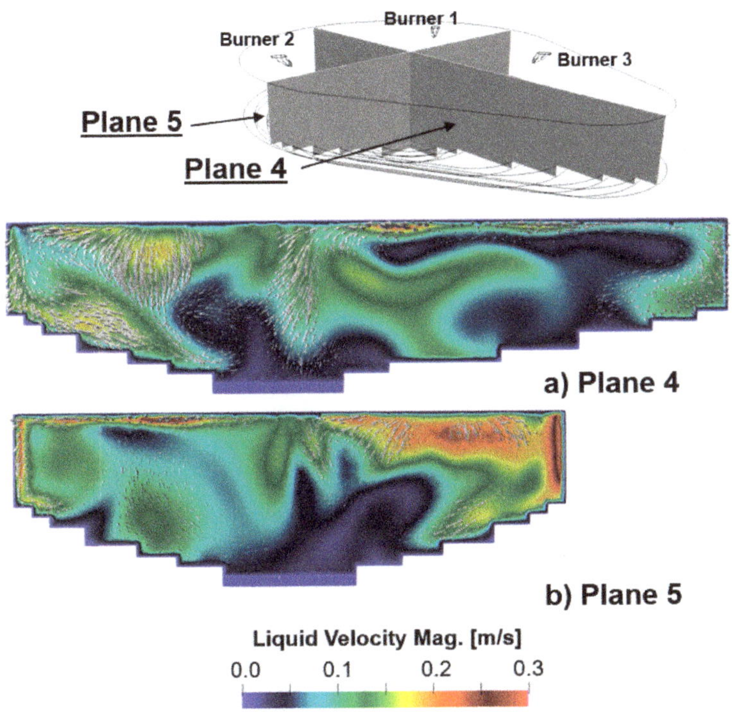

Figure 11. Velocity magnitude and flow vectors computed at (**a**) plane 4: vertical plane oriented towards the balcony region, and (**b**) plane 5: vertical plane normal to plane 4.

Figure 13 shows the instantaneous velocity components U_{liquid}, V_{liquid}, and W_{liquid} of the liquid steel on planes 1, 2, and 3 discussed earlier. The axis orientation is included at the bottom left of the figure. Figure 13a,b show that the flow intensity is larger on the left side of the planes, near burners 1 and 2. Figure 13b shows that U_{liquid} and V_{liquid} increase as the flow interacts with the end wall at the balcony. The W_{liquid} contours in Figure 13a show the flow structures produced vertically as a result of the oxygen injection. Namely, the oxygen injection produces the stirring of the liquid bath upwards. The W_{liquid} contours in Figure 13a also show negative velocities associated with the stirring, which determine the recirculating patterns seen in planes 4 and 5 in Figure 11. The overall intensity of flow vertical penetration decreases as we move into the liquid bath, and a much reduced W_{liquid} contour is observed at the furnace bottom in Figure 13c.

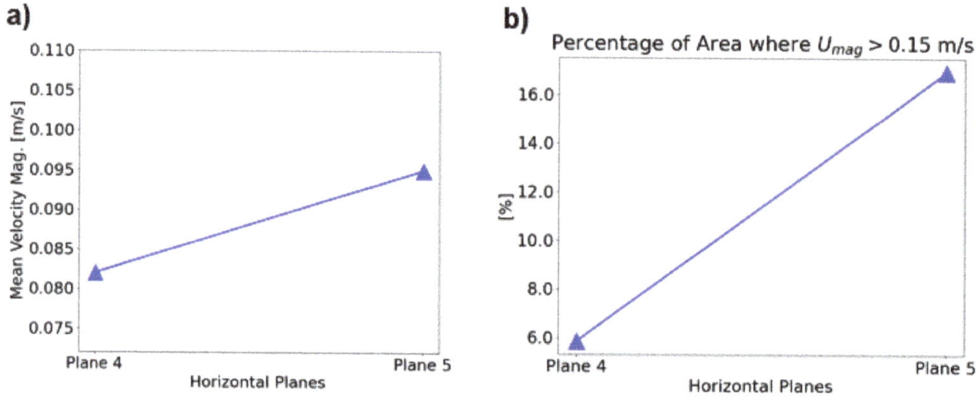

Figure 12. (**a**) Mean velocity magnitude, and (**b**) percentage of area where the velocity magnitude is larger than 0.15 m/s for the planes considered in Figure 11.

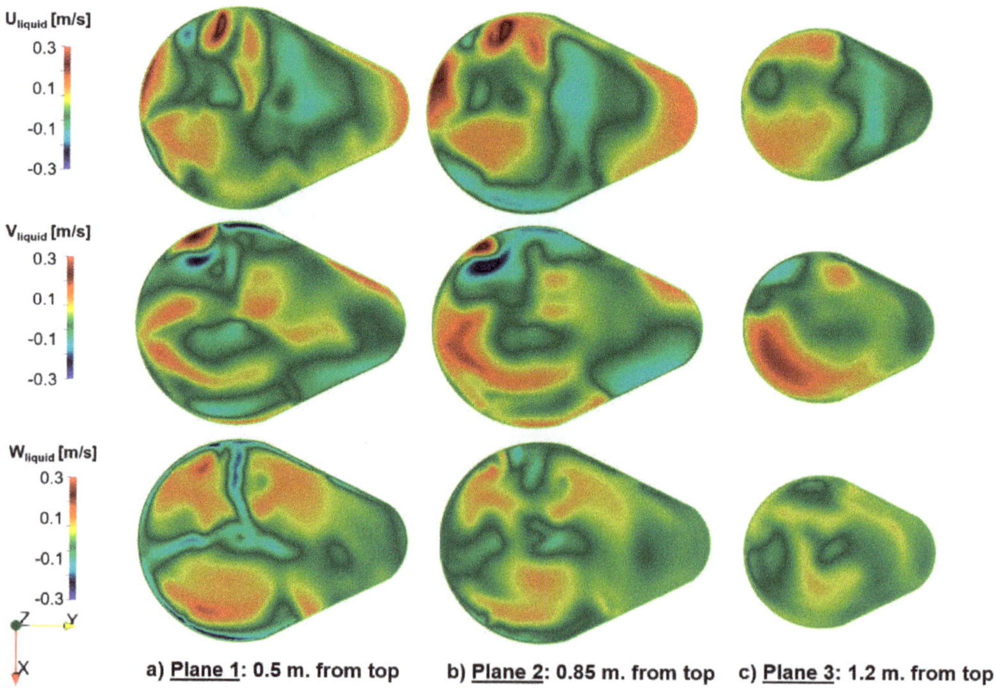

Figure 13. Velocity components computed at (**a**) plane 1: 0.5 m from bath surface, (**b**) plane 2: 0.85 m from bath surface, and (**c**) plane 3: 1.2 m from bath surface, in the baseline case.

3.3. Effect of Total Flow Rate

The effect of both increasing and decreasing the flow rate of the burners is explored in this section. Specifically, two additional cases are simulated where the flow rates at the burners are uniformly increased by 25% and decreased by 25%. These are listed in Table 3 as cases 2 and 3, respectively. It should be noted that coherent jet burners are designed to operate at 20–30% of their factory-set flow rate, as this helps maintain the stability of the flame, and preserves maximum oxygen delivery to the bath.

Table 3. Case conditions of the parametric study.

Cases	Coherent Jet Flow Rates (SCFM) (kg/s)				Stirring Energy (W/ton)
	Burner 1	Burner 2	Burner 3	Total Injection	
1	1000 (0.63)	1000 (0.63)	1000 (0.63)	3000 (1.88)	0.078
2	1250 (0.78)	1250 (0.78)	1250 (0.78)	3750 (2.34)	0.104
3	750 (0.47)	750 (0.47)	750 (0.47)	2250 (1.41)	0.130

Table 3 includes the stirring energy for cases 1–3. The stirring energy is computed based on the relation reported by Mazumdar and Guthrie [13], $\varepsilon = g.Q/\pi R^2$, where g is gravity acceleration, Q is the gas flow rate in Nm^3/s, and R is the averaged radius of the furnace. Since the EAF geometry is asymmetric in this study, R is defined as the semi-sum of long and short lengths of the liquid surface, leading to R = 6.28 m.

The effect of increasing and decreasing the flow rate in the burners is shown in Figure 14, where these scenarios, in addition to the baseline case, are compared at t = 600 s. Figure 14b shows that the increased flow rate intensifies the flow near the burners, and leads to the formation of a 'dead zone' at the center of the furnace, where the flow velocity is negligible. The flow vectors in Figure 14b show a significant increase in the liquid velocity near burners 2 and 3, and near the walls, as compared to the baseline case. Figure 14c shows the case with a decreased flow rate. Here, the flow velocities are reduced throughout the domain, and the dead zone regions are the largest among the three scenarios.

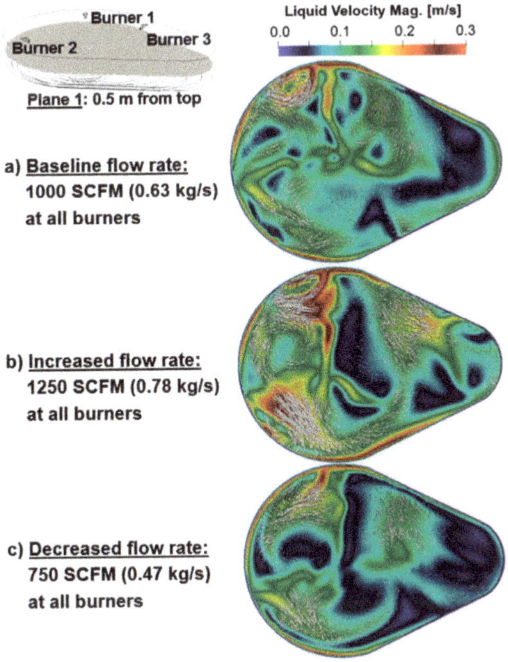

Figure 14. Velocity magnitude and flow vectors of (**a**) baseline case, (**b**) increased flow rate and (**c**) reduced flow rate, at t = 600 s.

In addition to the impact of the modified injection rates on the velocity magnitudes, changes in the injection rates also modify the flow patterns, reflected in the location of the recirculation zones and the formation of the dead zones. Figure 15 compares the three cases listed in Table 3 by computing the velocity magnitude on a plane taken along the centerline of the furnace, as shown at the top of the figure. Figure 15b shows that the increased flow rate case moves the recirculation region towards the left side of the plane, and makes the 'jet' type flow in the center of the domain (along the vertical direction) stronger than in the baseline case. The increased flow rate case strengthens the two vortices on the left side of the jet-type vertical flow. By contrast, when the flow rate is reduced in all the burners from 1000 to 750 SCFM, one vortex is observed near the left wall of the plane, and the flow penetration towards the bottom of the furnace weakens significantly (Figure 15c).

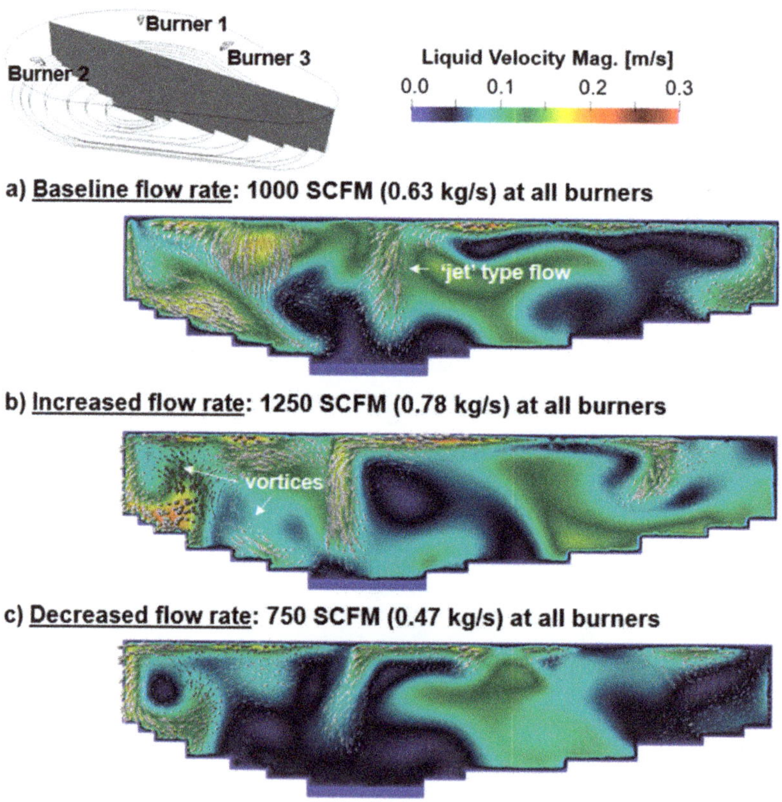

Figure 15. Velocity magnitude and flow vectors of (**a**) baseline case, (**b**) increased flow rate and (**c**) reduced flow rate, computed in a vertical plane.

Figure 16 shows regions in the furnace domain where the instantaneous velocity magnitude is larger than 0.15 m/s for the baseline case and for the cases with increased and reduced flow injection rates. These regions confirm the highly asymmetric distribution of the flow velocities, as the larger velocities are obtained in the half of the domain that contains burners 1 and 2. Figure 16b shows that by increasing the flow injection from 1000 to 1250 SCFM, the regions containing velocities above 0.15 m/s extend towards the balcony of the furnace.

Figure 17 helps to illustrate the impact of increasing and reducing the injection rate uniformly in all the burners. Figure 17a shows that the mean velocity magnitude increases by 10% when increasing the flow rate from 750 to 1000 SCFM, and by 23% when increasing the flow rate from 1000 to 1250 SCFM. However, the impact on the size of the region where the flow velocity is significant (i.e., larger than 0.15 m/s) seems rather exponential, as it increases by 25% when increasing the flow rate from 750 to 1000 SCFM, and 210% when increasing the flow rate up to 1250 SCFM. Therefore, increasing the flow rate has a larger impact on extending the region where the jets stir the bath rather than increasing the liquid steel velocity itself.

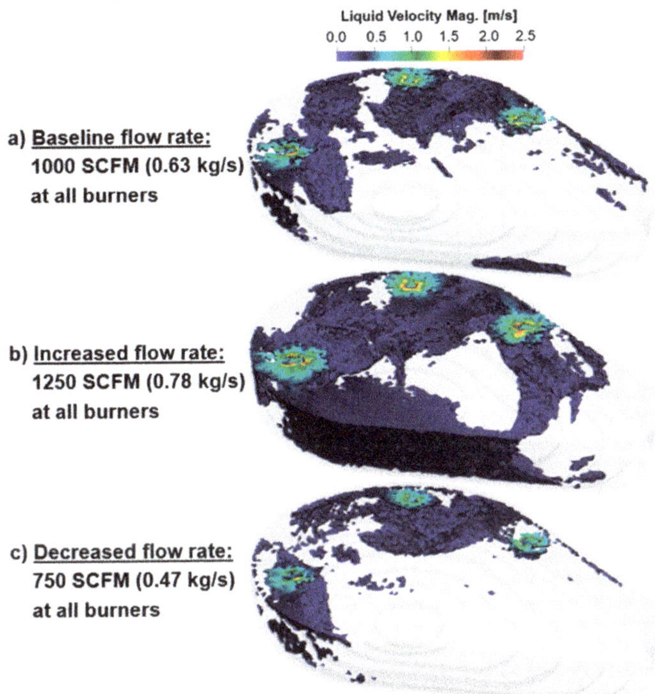

Figure 16. Regions where the velocity magnitude is larger than 0.15 m/s for (**a**) baseline case, (**b**) increased flow rate, and (**c**) reduced flow rate cases.

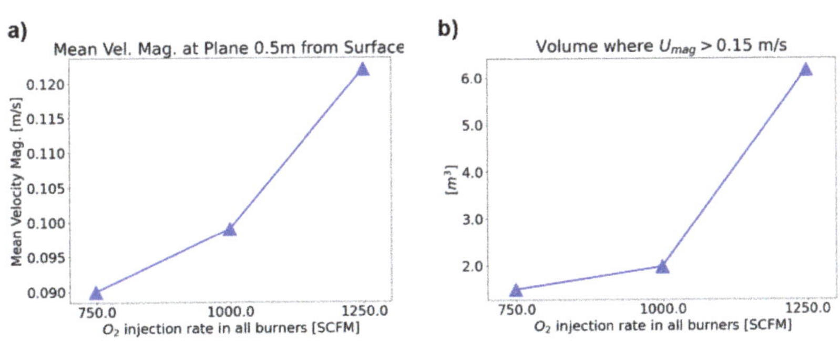

Figure 17. (**a**) Mean velocity magnitude, and (**b**) volume of regions where velocity magnitude is larger than 0.15 m/s for the baseline, increased, and reduced injection cases.

3.4. Effect of Non-Uniform Burner Flow Rates

The analysis performed in Sections 3.2 and 3.3 is extended here to include individual variations in the flow rate of the coherent jets. This leads to non-uniform injection rate scenarios. Three cases are considered in this study, which are designated as cases 4 through 6 in Table 4.

Table 4. Conditions of cases where the jet rates are varied individually.

Cases	Coherent Jet Flow Rates (SCFM) (kg/s)			
	Burner 1	Burner 2	Burner 3	Total Injection
4	750 (0.47)	1000 (0.63)	1250 (0.78)	3000 (1.88)
5	1250 (0.78)	750 (0.47)	1000 (0.63)	3000 (1.88)
6	1000 (0.63)	1250 (0.78)	750 (0.47)	3000 (1.88)

The contours of the liquid velocities and flow vectors of the non-uniform injection scenarios at t = 600 s are shown in Figure 18. In these cases, the flow rate injected by the three burners combined is the same, 3000 SCFM, which differs from cases 1–3 where the total injection rate changed. In cases 4–6, the flow pattern varies significantly depending on the location of the burner that is increased or decreased with respect to the baseline rate of 1000 SCFM. Figure 18b shows the case where the largest flow rates are at burners 2 and 3. In this case, the dead zone region is increased in the center of the furnace as compared to the baseline due to the stirring driven mainly by burners that are located opposite each other. By increasing the flow rate in burner 1 and reducing it in burner 2 (Figure 18c) the largest flow rates are injected near the top side of the plane, where burners 1 and 3 are located. This also increases the flow velocities near burner 2, even though the flow rate in this burner is reduced to 750 SCFM. In this case, the flow velocities are larger than in the baseline case, although a dead zone region is formed in the center of the domain in addition to the one formed at the balcony. Finally, by setting burners 1 and 2 to the largest flow rates while reducing the injection rate at burner 3 (Figure 18d), the flow velocities are intensified as well, but the increase in the flow velocities is mostly on the left side of the plane view, whereas the right side and balcony region are exposed to low liquid velocities.

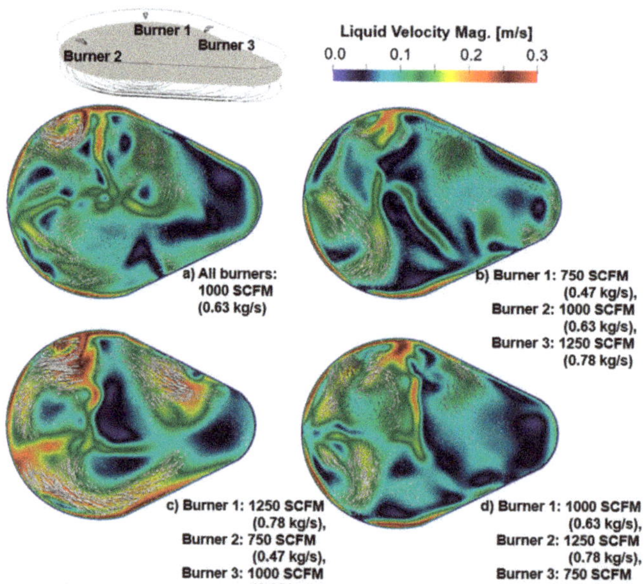

Figure 18. The velocity magnitude and flow vectors of (**a**) baseline case, and individually modified injection rate cases: (**b**) larger injection rates at burners across each other, (**c**) larger injection rates at one side of the furnace, and (**d**) larger injection rates at the front of the furnace.

Figure 19 compares the flow velocities of liquid steel on a plane taken along the furnace centerline, towards the balcony region. In all four cases, there is a 'jet'-type flow from the liquid surface toward the bottom of the furnace near the center of the plane. The location of this jet flow changes with the flow rate of the burners. In Figure 19a,d, the jet flow is seen closer to the center of the plane than in Figure 19b,c. In all cases, flow recirculation is observed to the left of the jet flow. The location and intensity of the jet flow are expected to have a significant impact on the mixing process, which will be analyzed in the next section.

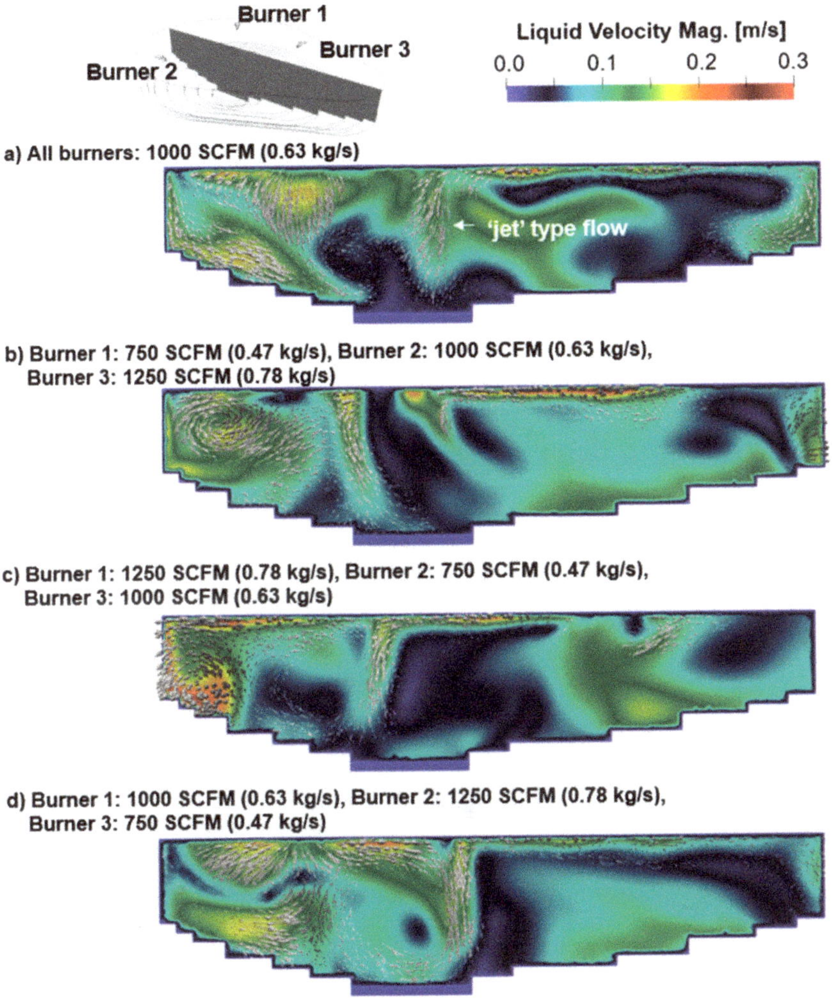

Figure 19. The velocity magnitude and flow vectors of (**a**) baseline case, and individually modified injection rate cases: (**b**) larger injection rates at burners across each other, (**c**) larger injection rates at one side of the furnace, and (**d**) larger injection rates at the front of the furnace, all computed on a vertical plane.

Figure 20 shows the regions where the liquid flow velocities are larger than 0.15 m/s for the cases listed in Table 4. As mentioned earlier, the total flow rate is the same for all the cases, but clearly the stirring intensity is not, as the injection rate in each of the burners is modified. The largest flow intensity is seen in Figure 20c, and a moderated intensification of the flow is seen in Figure 20d. In these two cases, the burners injecting the largest flow rates, 1000 and 1250 SCFM, are located next to each other. By contrast, in Figure 20b the largest flow rates are applied by burners opposite each other (burners 2 and 3), and no significant difference with respect to the baseline case is observed.

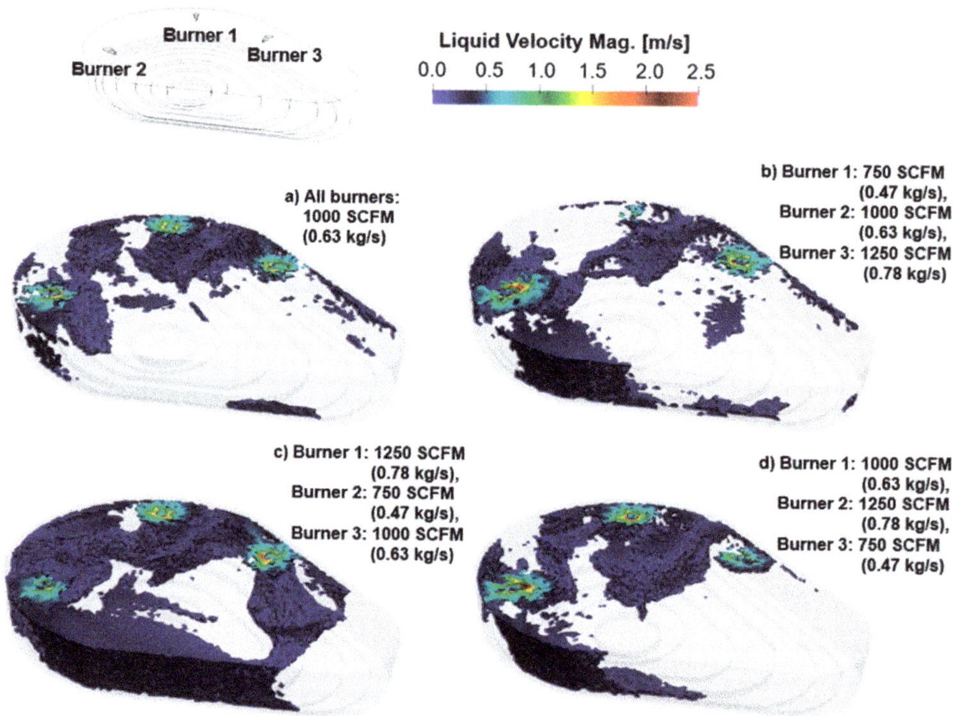

Figure 20. Regions where the velocity magnitude is larger than 0.15 m/s for (**a**) baseline case, and individually modified injection rate cases: (**b**) larger injection rates at burners across each other, (**c**) larger injection rates at one side of the furnace, and (**d**) larger injection rates at the front of the furnace.

Figure 21 shows the volume of the regions where the velocity magnitude is larger than 0.15 m/s for the baseline case and cases 4 through 6. These results show that the flow rate distribution in case 4 reduces the region of larger flow intensity obtained in the baseline case. Case 5 increases the large velocity region by 260% with respect to the baseline, whereas Case 6 increases the large velocity region by 15%.

The last set of results presented in this section corresponds to the W_{liquid} velocity computed at 0.5 m from the top of the liquid bath, for all six cases presented in this study (Figure 22). The W_{liquid} velocity is the velocity component that is perpendicular to the surface of the liquid bath, and it is expected to have a large impact on the reactions produced in the steel–slag interface and on the slag mixing during the actual operation of the EAF. Figure 22 shows three regions in each of the cases where W_{liquid} is positive (flow moving towards the bath surface). These regions correlate with the locations of the burners. Interestingly, the case with the reduced flow injection (750 SCFM, Figure 22a) shows larger W_{liquid} velocities than the baseline case and the case with increased flow injection (Figure 22b,c, respectively). These larger W_{liquid} velocities are observed near the walls, adjacent to the burners. However, Figure 22a also exhibits the larger regions with negligible W_{liquid} velocity among the three uniform injection cases.

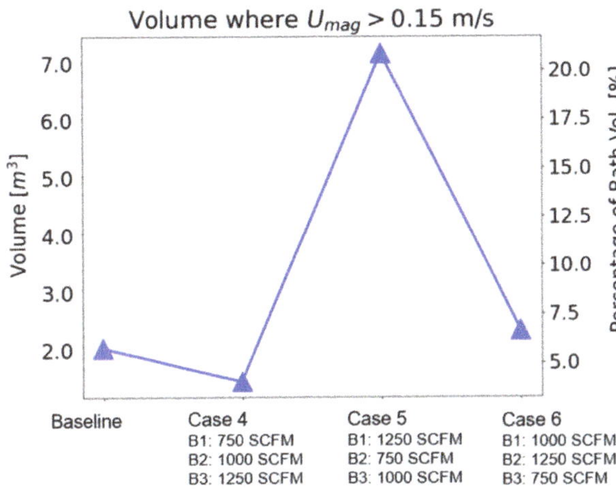

Figure 21. The volume of regions where the velocity magnitude is larger than 0.15 m/s for the baseline and the individually modified injection rate cases (volume of liquid bath is 34.3 m^3).

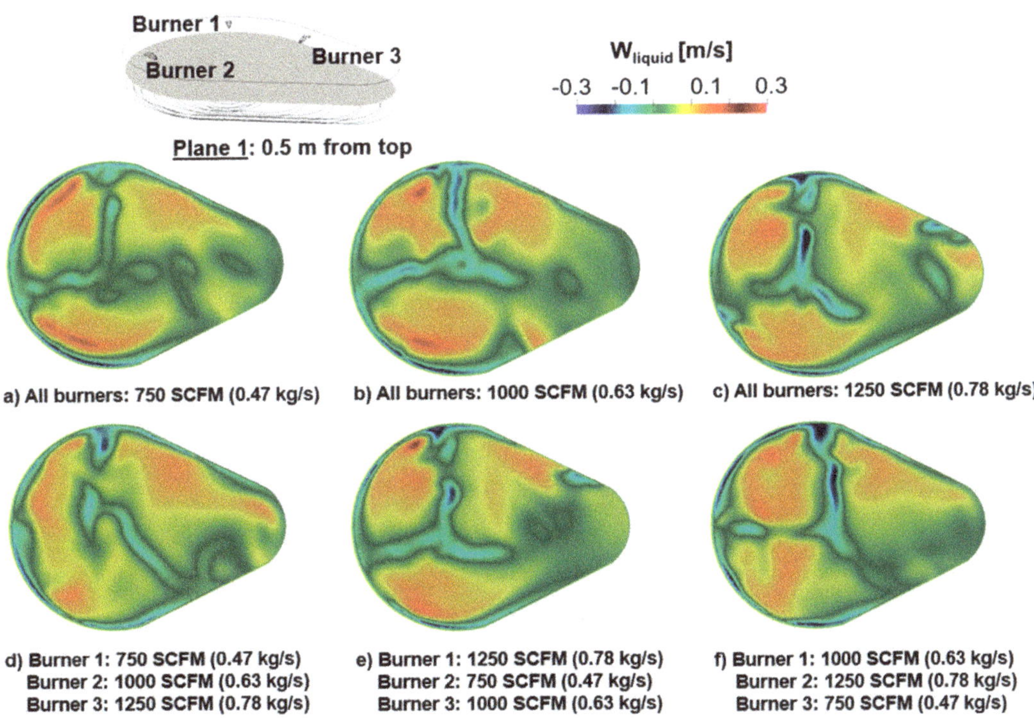

Figure 22. W_{liquid} velocities computed near the bath surface for cases (**a**) uniformly reduced injection rate, (**b**) baseline case, (**c**) uniformly increased injection rate, (**d**) larger injection rates at burners across each other, (**e**) larger injection rates at one side of the furnace, and (**f**) larger injection rates at the front of the furnace. Positive W_{liquid} shows flow towards the bath surface.

Figure 22d–f show the cases with non-uniform injection rates. Among these results, Figure 22d shows the case with the smallest region containing zero or near zero W_{liquid} velocities. However, the largest W_{liquid} velocities are seen in Figure 22e. As mentioned earlier, the W_{liquid} velocity component is expected to impact the reactions near the steel/slag interface. In this paper, however, neither the reactions nor the presence of slag are considered, and the study is focused on the flow features resulting under different injection rates and the way this impacts the mixing rate. In the next section, the flow velocity variations and corresponding mixing rates are studied in more detail for the different injection scenarios shown in cases 1–6.

Figures 23 and 24 complement the previous set of results by showing the regions where low velocities are produced. These results correspond to regions where the velocity magnitude is less than 0.05 m/s. The dead zones impact the mixing rate of the liquid bath as these regions are expected to take longer to stir, which will lead to the formation of velocity (and concentration) gradients in the bath. Figure 23 shows dead zones mainly in the central region and on the balcony of the furnace, for all the cases. The dead zones on the balcony are reduced in Figure 23b,c, as the flow rate of burner 3 located near this region is significant. The dead zones in the center of the domain are reduced in Figure 23d, as a result of the large flow rates imposed by burners 1 and 2, located nearby.

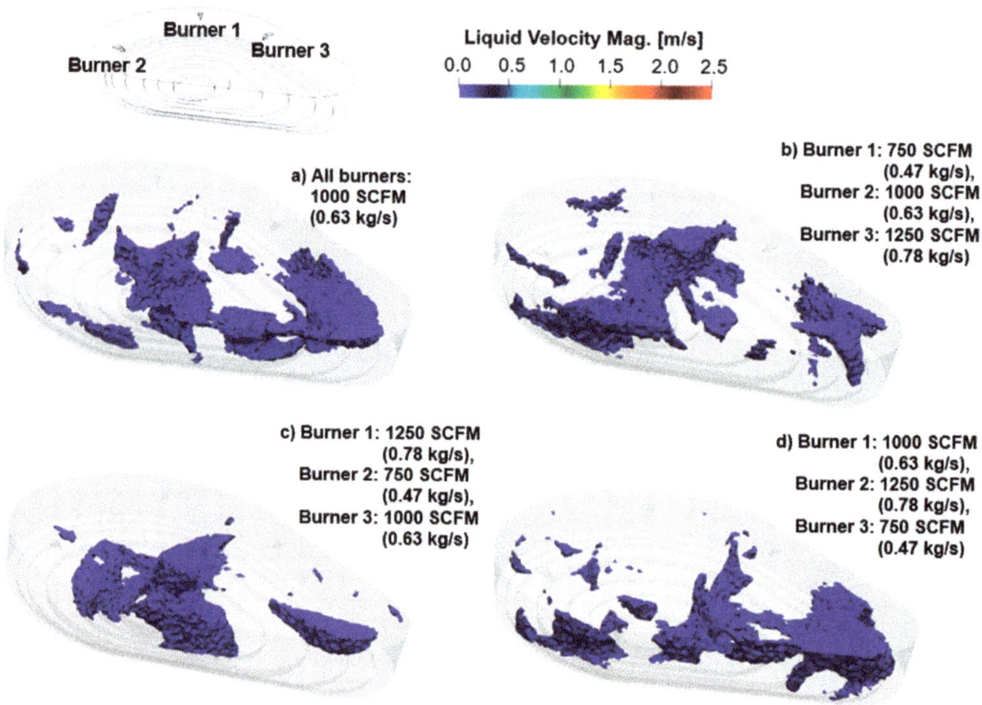

Figure 23. Regions where the velocity magnitude is less than 0.05 m/s for (**a**) baseline case, and individually modified injection rate cases: (**b**) larger injection rates at burners across each other, (**c**) larger injection rates at one side of the furnace, and (**d**) larger injection rates at the front of the furnace.

Figure 24 shows the size of the bath regions shown in Figure 23. It is shown that the non-uniform injection rates of the burners reduce the 'dead zones' size with respect to the baseline case. This reduction is minimal for case 5, but ~20% for cases 4 and 6. As mentioned earlier, the reduction in the size of the dead zones should have a positive impact on the efficiency of the mixing rate, which will be analyzed in the next section.

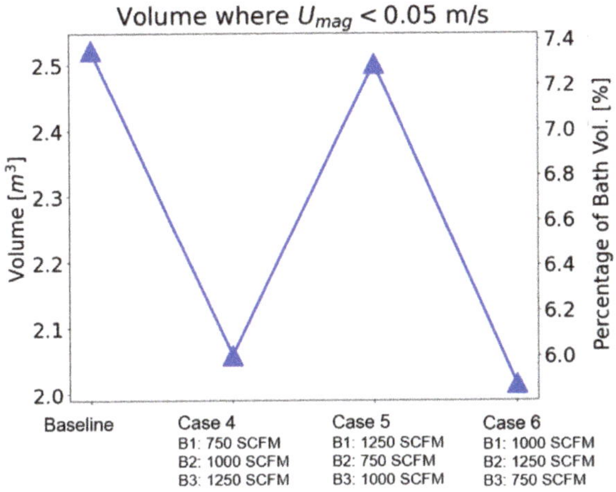

Figure 24. The volume of regions where velocity magnitude is less than 0.05 m/s for the baseline case and the individually modified injection rate cases (volume of liquid bath is 34.3 m^3).

4. Discussion

In order to quantify the effect of injection rates on bath mixing, two parameters are defined: the mixing time, which is the time needed by any species to blend into the generated flow, and the standard deviation of the fluid flow velocity, which determines the homogeneity of the velocity field in the liquid bath.

The mixing time is computed by introducing a passive tracer in the generated flow, with a mass fraction $Y_{tracer} = 1.0$. The uniformity index, which represents how a specified field variable varies over the domain, is used to calculate the mixing time. Specifically, the uniformity index is calculated as follows:

$$y_a = \frac{1}{V} \sum_{i=1}^{n} \varnothing_i \, |V_i| \tag{12}$$

where $\overline{\varnothing_i}$ is the local value of the field variable. In this study, $\overline{\varnothing_i}$ is given by Y_{tracer}. According to this definition, a uniformity index of 1 indicates a uniform concentration of the passive tracer throughout the domain.

The variation in the uniformity index with time can be calculated locally on selected planes along the simulation. This requires computing the uniformity index on a per area basis rather than per volume as given by Equation (12). Figure 25 shows the uniformity variation with the time of selected planes for the baseline case (1000 SCFM in all the burners). The planes where the uniformity index is computed are shown on the right side of the figure. The results in Figure 25 show how the uniformity index varies in time with the selected location, and how the uniformity index value eventually converges to ~1 at a similar time for all planes, around 640 s. In the results to be discussed next, the mixing time in each of the cases is determined by the time when the uniformity index calculated on a per volume basis (by using Equation (12)) reaches 0.95.

Figure 26 shows the mixing time for cases 1 (baseline, 1000 SCFM in all burners), 2 (1250 SCFM in all burners), and 3 (750 SCFM in all burners). This illustrates the effect of increasing and reducing the flow rate of the coherent jet on the mixing process. Namely, a 25% increase in the flow rate leads to a reduction of 6.7% in the mixing time, whereas by reducing the uniform flow rate by 25%, the mixing time increases by 10.9%.

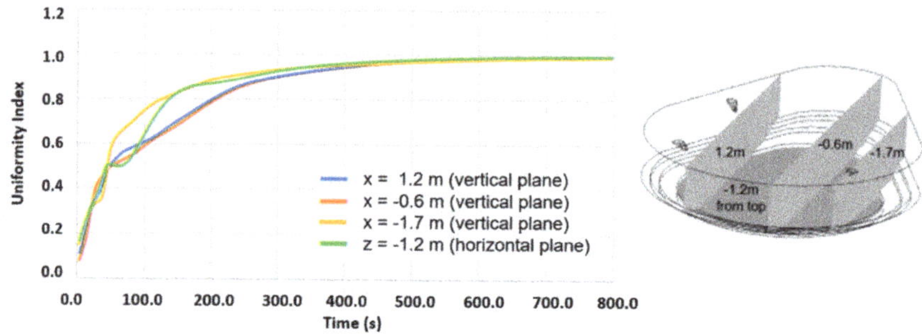

Figure 25. Local variations in the uniformity index with time for the baseline case.

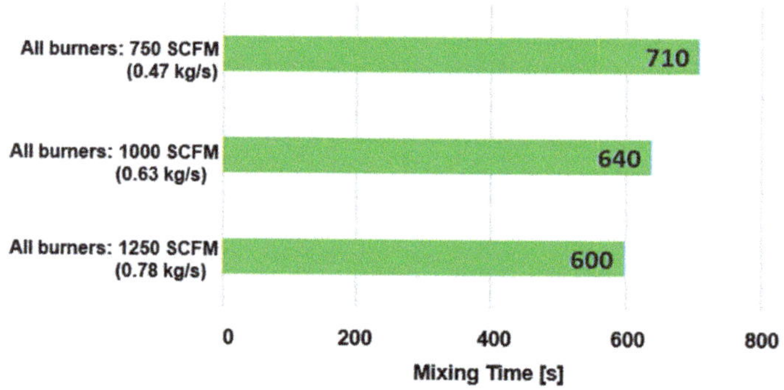

Figure 26. Mixing time for cases with uniform flow injection rates.

Figure 27 compares the mixing time and standard deviation of the baseline case and the cases with non-uniform injection rates (cases 4 through 6 in Table 4). The results show that the non-uniform injection cases reduce the mixing time obtained in the baseline case (green bars in Figure 25). This reduction is correlated with an increase in the standard deviation of the flow velocity with respect to the baseline case (yellow bars in Figure 27).

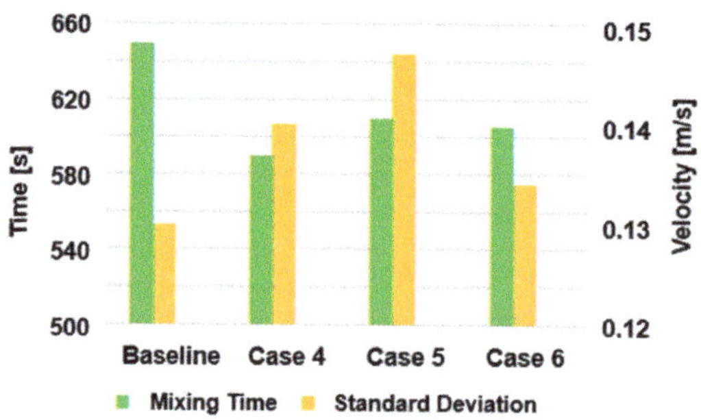

Figure 27. Mixing time and standard deviation of velocity for the baseline case and the cases with individually modified flow injection rates.

Figure 27 also shows that the shortest mixing time does not correlate with the largest standard deviation of the flow velocity. Specifically, case 4 is the most optimal as it reduces the mixing time the most: 10.2% with respect to the baseline. Figures 18b and 19b show that case 4 (burner 1 = 750 SCFM, burner 2 = 1000 SCFM, and burner 3 = 1250 SCFM) prevented the formation of dead zones in the balcony region, and was able to continue stirring the flow in the center of the domain. Figures 23 and 24 showed that cases 4 and 6 led to the largest reduction in dead zones in the bath, which prevented the mixing of the flow. Figure 27 also shows that the standard deviation in case 4 is larger than in case 6, which contributes to reducing the mixing time. Although case 4 did not lead to the most intense flow field, the injection distribution in this case led to the most optimal mixing scenario by addressing the stirring of the central region of the furnace and the balcony regions which, based on the multiple scenarios investigated, are the most difficult to stir.

It should be noted that by modifying the flow field it is possible to impact the consumption of electrodes and refractories, as a result of the shear stresses developed in the flow. Specifically, it is expected that a more intense flow field will lead to better mixing, reducing the refining time, but will also increase the shear stress between the flow and surfaces in the EAF, increasing the consumption of both refractories and electrodes.

5. Conclusions

In this study, a CFD platform was used to analyze the stirring of molten steel driven by oxygen injection in an industrial-scale EAF. CFD was used to perform non-reacting, three-dimensional, transient simulations of the liquid bath as it interacts with three coherent jets operating in lance mode. The coherent jets are distributed in an asymmetric manner in the furnace.

A baseline simulation is set based on the operating conditions of a typical industrial EAF. In this case, the coherent jets inject oxygen at 1000 SCFM in all three burners. The results show that large fluid velocities near the burners drop significantly as the flow interacts with walls and fluids injected from adjacent burners. The flow pattern is highly non-uniform due to the geometry of the furnace and the asymmetric distribution of the injectors. In the baseline case, flow recirculation develops near the burners, in particular near burner 1. Also, 'dead zones' develop on the balcony of the furnace, at the wall opposite burner 3, and near the center of the domain. Analysis performed in vertical planes shows large velocities near burners 1 and 2 (front of the furnace) and vertical recirculation also developing in this region.

The effect of injection rate on the stirring process is first studied by increasing and decreasing the injection rate in all burners with respect to the baseline case. Namely, two injection cases were considered: 750 SCFM and 1250 SCFM in all burners. The increased flow rate strengthens the vortices formed in the vertical (center) plane of the furnace, whereas the reduced flow reduces the number of vortices developing in this direction. The impact of injection rates on the flow mixing is further studied by simulating three cases where the injection rate is modified in each burner, but the total flow rate is kept at 3000 SCFM. The results show that by keeping the largest injection rates at burners 2 and 3, opposite each other, it is possible to reduce the dead zones in the balcony while maintaining the non-uniformity of the flow velocities. The individual injection rates modify the formation of the 'jet'-type flow in the vertical direction, moving the jet flow towards the front of the furnace when the injection rate of any of the front burners (burners 1 and 2) is decreased.

The flow analysis of these cases is quantified by introducing two variables: the mixing time and the standard deviation of the flow velocity. The mixing time was computed based on the concentration of a passive tracer injected into the flow field. The results of the uniform injection cases show that mixing time reduces as the injection rate set in all the burners is increased. Namely, a 25% increase in the flow rate leads to a 6.7% reduction, whereas by reducing the uniform flow rate by 25%, the mixing time increases by 10.9%. Moreover, the non-uniform injection rate cases improve the mixing obtained in the baseline case. Namely, the mixing time declined by 10.2% when the largest flow rates were set to coherent jets located opposite each other in the furnace (burner 1 = 750 SCFM, burner 2 = 1000 SCFM, and burner 3 = 1250 SCFM). The results also showed that the non-uniform injection cases increased the standard deviation of the flow velocities in the furnace with respect to the baseline, which led to an overall improvement in the stirring process and a reduction in the time needed to reach a homogenous concentration field.

Author Contributions: Conceptualization, O.U., N.B., T.O. and B.K.; methodology, O.U. and N.B.; software, N.B.; formal analysis, O.U. and N.B.; investigation, O.U. and N.B.; resources, T.O. and C.Q.Z.; data curation, O.U. and N.B.; writing—original draft preparation, O.U., N.B. and B.K.; writing—review and editing, O.U. and B.K.; visualization, O.U. and N.B.; supervision, T.O. and C.Q.Z.; project administration, T.O. and C.Q.Z.; funding acquisition, T.O. and C.Q.Z. All authors have read and agreed to the published version of the manuscript.

Funding: This research received no funding.

Data Availability Statement: Simulation data for generic geometries and operating conditions may be made available upon request of the authors. Simulation data pertaining to industrial operations is protected by confidentiality and may only be shared with the permission of industrial partners within the Steel Manufacturing Simulation and Visualization Consortium.

Acknowledgments: The authors would like to thank the members of the Steel Manufacturing Simulation and Visualization Consortium (SMSVC) for their support and consultation on this research. The authors would also like to thank the staff and students at Purdue University Northwest's Center for Innovation through Visualization and Simulation (CIVS) for their support and for the resources to conduct this research. Finally, the authors would like to acknowledge the great assistance and input on EAF CFD modeling provided by Yuchao Chen.

Conflicts of Interest: On behalf of all authors, the corresponding author states that there are no conflicts of interest.

References

1. Fan, Z.; Friedmann, S.J. Low-carbon production of iron and steel: Technology options, economic assessment, and policy. *Joule* **2021**, *5*, 829–862. [CrossRef]
2. Puschmann, T.; Hoffmann, C.H.; Khmarskyi, V. How Green FinTech Can Alleviate the Impact of Climate Change—The Case of Switzerland. *Sustainability* **2020**, *12*, 10691. [CrossRef]
3. Conejo, A.N.; Yan, Z. Electric Arc Furnace Stirring: A Review. *Steel Res. Int.* **2023**, *94*, 2200864. [CrossRef]
4. Dreyfus, L. Electric Furnace. U.S. Patent 2,256,518, 1941.
5. Teng, L.; Meador, M.; Ljungqvist, P. Application of New Generation Electromagnetic Stirring in Electric Arc Furnace. *Steel Res. Int.* **2017**, *88*, 1600202. [CrossRef]
6. Chen, Y.; Silaen, A.K.; Zhou, C.Q. 3D Integrated Modeling of Supersonic Coherent Jet Penetration and Decarburization in EAF Refining Process. *Processes* **2020**, *8*, 700. [CrossRef]
7. Jayaraj, S.; Kang, S.; Suh, Y.K. A review on the analysis and experiment of fluid flow and mixing in micro-channels. *J. Mech. Sci. Tech.* **2007**, *21*, 536–548. [CrossRef]
8. Kang, T.G.; Anderson, P.D. The Effect of Inertia on the Flow and Mixing Characteristics of a Chaotic Serpentine Mixer. *Micromachines* **2014**, *5*, 1270–1286. [CrossRef]
9. Liu, R.H.; Stremler, M.A.; Sharp, K.V.; Olsen, M.G.; Santiago, J.G.; Adrian, R.J.; Aref, H.; Beebe, D.J. Passive mixing in a three-dimensional serpentine microchannel. *J. Microelectromech. Syst.* **2000**, *9*, 190–197. [CrossRef]
10. Lang, E.; Drtina, P.; Streiff, F.; Fleischli, M. Numerical simulation of the fluid flow and the mixing process in a static mixer. *Int. J. Heat Mass Transf.* **1995**, *38*, 2239–2250. [CrossRef]
11. Li, Y.; Lou, W.T.; Zhu, M.Y. Numerical simulation of gas and liquid flow in steelmaking converter with top and bottom combined blowing. *Ironmak. Steelmak.* **2013**, *40*, 505–514. [CrossRef]
12. Duan, Y.B.; Wei, J.H. Influence of rotating of side blowing gas jets on fluid mixing characteristics in a 120t AOD bath with side and top combined blowing. *Shanghai Met.* **2007**, *29*, 31–36.
13. Mazumdar, D.; Guthrie, R.I. Mixing models for gas stirred metallurgical reactors. *Metall. Mater. Trans. B* **1986**, *17*, 725–733. [CrossRef]
14. Zhu, M.Y.; Inomoto, T.; Sawada, I.; Hsiao, T.C. Fluid fow and mixing phenomena in the ladle stirred by argon through multi-tuyere. *ISIJ Int.* **1995**, *35*, 472–479. [CrossRef]
15. Amaro-Villeda, A.M.; Ramirez-Argaez, M.A.; Conejo, A.N. Effect of Slag Properties on Mixing Phenomena in Gas-stirred Ladles by Physical Modeling. *ISIJ Int.* **2014**, *54*, 1–8. [CrossRef]
16. Cheng, R.; Zhang, L.; Yin, Y.; Zhang, J. Effect of Side Blowing on Fluid Flow and Mixing Phenomenon in Gas-Stirred Ladle. *Metals* **2021**, *11*, 369. [CrossRef]
17. Li, B. Fluid flow and mixing process in a bottom stirring electrical arc furnace with multi-plug. *ISIJ Int.* **2000**, *40*, 863–869. [CrossRef]
18. Tang, G.; Chen, Y.; Silaen, A.K.; Krotov, Y.; Riley, M.F.; Zhou, C.Q. Investigation on coherent jet potential core length in an electric arc furnace. *Steel Res. Int.* **2019**, *90*, 1800381. [CrossRef]
19. Sano, M.; Mori, K. Fluid flow and mixing characteristics in a gas-stirred molten metal bath. *Trans. Iron Steel Inst. Jpn.* **1983**, *23*, 169–175. [CrossRef]

20. Banks, R.B.; Chandrasekhara, D.V. Experimental investigation of the penetration of a high-velocity gas jet through a liquid surface. *J. Fluid Mech.* **1963**, *15*, 13–34. [CrossRef]
21. Ishikawa, H.; Mizoguchi, S.; Segawa, K. A model study on jet penetration and slopping in the LD converter. *ISIJ Int.* **1972**, *58*, 76–84.
22. Gonzalez, O.J.P.; Ramírez-Argáez, M.A.; Conejo, A.N. Effect of arc length on fluid flow and mixing phenomena in AC electric arc furnaces. *ISIJ Int.* **2010**, *50*, 1–8. [CrossRef]

metals

MDPI

Article

Biocoke Thermochemical Properties for Foamy Slag Formations in Electric Arc Furnace Steelmaking

Lina Kieush [1,*], Johannes Schenk [2], Andrii Koveria [3] and Andrii Hrubiak [4]

[1] K1-MET GmbH, 8700 Leoben, Austria
[2] Chair of Ferrous Metallurgy, Montanuniversitaet Leoben, 8700 Leoben, Austria
[3] Department of Chemistry, Dnipro University of Technology, 49005 Dnipro, Ukraine
[4] G.V. Kurdyumov Institute for Metal Physics of the National Academy of Sciences of Ukraine, 02000 Kyiv, Ukraine
* Correspondence: lina.kieush@k1-met.com; Tel.: +43-38420422278

Abstract: This paper is devoted to studying the thermochemical properties of carbon sources (laboratory-scale conventional coke, biocoke with 5 wt.%, and 10 wt.% wood pellet additions) and the influence of these properties on foamy slag formations at 1600 °C. Thermogravimetric analysis (TGA) conducted under air unveiled differences in mass loss among carbon sources, showing an increasing order of coke < biocoke with 5 wt.% wood pellets < biocoke with 10 wt.% wood pellets. The Coats–Redfern method was used to calculate and reveal distinct activation energies among these carbon sources. Slag foaming tests performed using biocoke samples resulted in stable foam formation, indicating the potential for biocoke as a carbon source to replace those conventionally used for this process. Slag foaming characters for biocoke with 5 wt.% wood pellets were improved more than coke. Using biocoke with 10 wt.% wood pellets was marginally worse than coke. On the one hand, for biocoke with 5 wt.% wood pellets, due to increased reactivity, the foaming time was reduced, but it was sufficient and optimal for slag foaming. Conversely, biocoke with 10 wt.% wood pellets reduced foaming time, proving insufficient and limiting the continuity of the foaming. This study highlights that thermochemical properties play a significant role, but comprehensive assessment should consider multiple parameters when evaluating the suitability of unconventional carbon sources for slag foaming applications.

Keywords: coke; biocoke; thermochemical properties; foaming; slag; thermogravimetric analysis

1. Introduction

As it is imperative to reduce carbon emissions and environmental impact, there is a growing focus on exploring novel strategies to improve the efficiency and ecological viability of the electric arc furnace (EAF) steelmaking process [1]. Slag foaming, an important process of EAF steelmaking, hinges on injecting oxygen and carbon sources into the molten metal bath and slag layer, respectively. This dual-injection approach serves the dual purpose of oxidizing impurities while fostering the formation of a stable and effective foamy slag [2]. Precisely, the carbon source reacts with the FeO, producing CO, which primarily produces foaming. The chemical reactions [3] generate CO/CO_2 gases, forming bubbles within the slag, thereby creating foam. Foamy slags serve to increase surface areas, facilitating multi-phase reactions. This enhances reaction kinetics, heat transfer, and energy efficiency while reducing electrode consumption [4–7]. Conventional carbon sources (anthracite, metallurgical coke, calcined petroleum coke, and graphite) applied are hardly dispensable in this process [8]. One such solution gaining prominence is the incorporation of biocoke [9] as a carbon source, offering a pragmatic and realistic approach to address the challenges of both performance and the environment. Biocoke can be used in slag foaming where the tolerance for reactivity and strength is not strict [10–12]. In addition to reactivity and strength, the thermochemical properties of carbon materials are important factors to consider when studying the possibility of using these materials for metallurgical processes [13–15]. This is especially important when the goal is to replace traditional ones with auxiliary sources, as it can determine the applicability and suitability of a particular process.

Thermogravimetric analysis (TGA) is a tool employed to investigate the thermal decomposition and combustion of materials, and it has found significant utility in the study of carbon material combustion processes [16]. Coke combustion is a complex process involving various chemical and physical processes. This includes the release of volatile matters, the ignition and continuous burning of fixed carbon, and the creation of ash residues. Each stage has its specific kinetics, which TGA can precisely investigate. TGA affords a precise means for quantifying mass variations as a function of temperature elevation, thereby yielding crucial insights into the quantification of weight loss attributed to the evolution of volatiles and the identification of the temperature thresholds for ignition and combustion. Furthermore, TGA enables the examination of reaction kinetics, facilitating the elucidation of reaction mechanisms and the evaluation of kinetic parameters.

Indeed, the non-isothermal TGA method to study the coke combustion characteristics in a pure oxygen atmosphere has been used by Yang et al. [17]. It was concluded that the change in activation energy E_a strongly depends on the conversion rate, reflecting the combustion process's complexity. Moreover, combustion characteristics and kinetics of high- and low-reactivity metallurgical cokes under air atmosphere were investigated by Qin et al. [18]. In this research, the coke combustion E_a was calculated using the Coats–Redfern method, as well as the Flynn–Wall–Ozawa (FWO) and Vyazovkin methods. The general tendency was that low-reactivity coke's E_a was more than that of high-reactivity coke. Additionally, the ignition and burnout temperatures were increased for low-reactivity coke, and the thermal stability of low-reactivity coke was improved compared to high-reactivity coke.

It is worth noting that practically no such studies have been carried out for biocoke. Although, considering the interest in using biomass or biochar, the thermochemical properties of these materials have been investigated to evaluate their applicability for EAF steelmaking. Fidalgo et al. [19] studied the thermochemical properties and reactivities of biochar derived from grape and pumpkin seeds to assess their suitability in EAF steelmaking. It was found that TGA analyses have revealed that grape seed char presents higher gasification and combustion reactivities and slower release of volatiles in comparison to the other samples, which may improve supply heat and sustain reactions.

In turn, Cardarelli et al. [20] studied the thermal effects of the combustion of pretreated biomasses via pyrolysis, torrefaction, and hydrothermal carbonization compared to anthracite coal, which were modeled for EAF application. The impact of the combustion reaction on the temperature distribution inside the EAF and the influence of intermediate gas release was analyzed. The results showed that using biochar instead of fossil coal in the EAF steelmaking process did not involve significant negative differences.

Specifying thermochemical parameters (thresholds of ignition temperature and burnout temperature; mass loss) can be of interest to substantiate the effect of carbon sources on foaming behavior, among other well-studied properties such as ash content and composition, volatile matters, fixed carbon, and microstructural features of carbon sources. Many researchers have investigated different slag foaming and carbon/slag interactions during this process [21]. However, since this process is a complex and unstable phenomenon, many factors can usually influence optimal achievement in slag foaming [22]. In studying slag foaming as a function of different carbon sources, in this case, it is quite challenging to compare the results obtained since the properties of one type of carbon material may differ, taking into account even equal test conditions.

However, some major tendencies should be mentioned:

– The value of fixed carbon, ash value, and composition are important in carbon source interactions with molten slag and slag foaming behavior [23].

- The value of volatile matters can also contribute to slag foaming behavior, but it is less effective than the gas from the chemical reaction [24].
- Carbon microstructure is an important factor worth considering [8].
- Wetting capacity, surface roughness, and chemical reactivity of carbon sources are important parameters influencing this process [8].

Among the array of carbon sources employed for slag foaming, biocoke could be considered an adequate candidate since, on the one hand, it allows for the reduction in coal consumption necessary to produce biocoke by replacing it with biomass/biochar; in addition, on the other hand, it is a source of carbon that complies with the required reactions [25] for foaming slag formation. Slag foaming is quite a complex and dynamic process affected by the following factors: type of carbon sources and their properties, composition of slag and its properties, test conditions, carbon/slag interfacial reaction kinetics, gas generation, etc.

Thus, this study focuses on thermogravimetric evaluation to examine the combustion characteristics and kinetics of coke and two biocoke samples. These thermochemical characteristics are expected to explain the influence of carbon sources during the foaming process and provide information for the slag foaming operation. The research undertaken is oriented toward defending the consequential impact of unconventional carbon sources on slag foaming formation within the confines of controlled experimental settings in each test.

2. Materials and Methods

The slag was obtained from powders characterized by specific chemical compositions, including iron (II) oxide, denoted as FeO; calcium oxide, denoted as CaO (≥96%); silica, denoted as SiO_2 (≥99%); magnesium oxide, denoted as MgO (≥98%); and aluminum oxide, denoted as Al_2O_3 (≥99%). The precise compositions are provided in Table 1. The oxide components and the target compositions are based on the primary chemical compounds in EAF slags [26]. This composition aims to induce slag foaming.

Table 1. Slag composition.

Sample	FeO, wt.%	CaO, wt.%	SiO_2, wt.%	MgO, wt.%	Al_2O_3, wt.%	Total	B2	B3
	29.0	35.0	17.0	10.0	9.0	100.0	2.0	1.3

B2, basicity CaO/SiO_2; B3, basicity $CaO/(SiO_2 + Al_2O_3)$.

For the production of laboratory-scale coke or biocoke specimens, 2 kg of a coal blend (for coke production) or a mixture of 95 wt.% coal and 5 wt.% wood pellets (for the biocoke production, referred to as BC5) or a mixture comprising 90 wt.% coal and 10 wt.% wood pellets (for the production of biocoke, referred to as BC10) were prepared. Afterward, a retort was introduced into a preheated electric shaft-type laboratory furnace, initially maintained at 850 °C. Following this initial preheating phase, the temperature was further elevated to 1000 °C and held for 75 min. The coking process occurred without oxygen access. During the coking process, volatile matters were ejected from the retort and combusted. Subsequently, the retort was withdrawn from the furnace, subjected to water cooling, and cooled under ambient air conditions until it reached room temperature.

The main characters of carbon sources are shown in Table 2. The proximate and elemental analysis for carbon sources were carried out according to ASTM D3172-13 [27]. Based on the obtained values for volatile matters and sulfur for the resulting coke and biocokes, they may not meet the requirements as fuel and reducing agents for blast furnace purposes [28]; however, if the fixed carbon content is also considered, coke and biocokes can be used, for example, for slag foaming in EAF steelmaking.

Table 2. Characters of carbon sources.

Characters	Coke	BC5	BC10
	Proximate analysis, wt.%		
M	0.77	0.66	0.65
VM (db)	1.46	1.40	1.42
VM (daf)	1.48	1.42	1.44
Ash (db)	11.1	10.8	10.5
C_{fix} (db)	87.4	87.8	88.1
	Higher heating value, MJ/kg		
HHV	29.57	29.59	29.63
	Elemental analysis, wt.%		
S (db)	1.36	1.28	1.20
C (db)	84.51	84.74	85.05
H (db)	0.30	0.27	0.22
N (db)	1.10	1.06	1.02
* Others, mainly O (db)	1.63	1.85	2.01
	Characteristic of microstructural parameter, nm		
d_{002}	0.36	0.35	0.35
L_a	3.60	2.98	3.04

M, moisture; VM, volatile matters; C, carbon; H, hydrogen; N, nitrogen; S, sulfur; db, dry basis; daf, dry ash-free basis; VM(daf) = VM(db)·100/(100 − Ash(db,%)); C_{fix}, wt.% = 100 − (wt.% VM(db) − wt.% Ash(db)); calculated by HHV = 0.3491 · C + 1.1783 · H + 0.1005 · S − 0.0151 · N − 0.1034 · O − 0.0211 · Ash. * Calculated by difference; d_{002}, aromatic planes of carbon crystallites; L_a, carbon crystallite width.

In preparation for the foaming experiments, all three carbon sources were subjected to particle size crushing to achieve a particle size range between 0.5 and 1.0 mm. The stoichiometric required amount of carbon source based on fixed carbon was calculated by considering the weight of the slag, the quantity of FeO within it, and the specific amount of fixed carbon needed for the FeO + C = Fe + CO reduction reaction and is presented in Table 3.

Table 3. Amount of carbon source based on fixed carbon for required reduction.

Sample	Coke, g	BC5, g	BC10, g
Stoichiometrically required amount	5.53	5.50	5.49

The slag foaming was conducted using an induction high-temperature furnace, the MU-900 (manufactured by Indutherm Erwärmungsanlagen GmbH, Walzbachtal, Germany), and involved the use of three carbon sources: conventional laboratory-scale coke (utilized as a reference), BC5, and BC10. Prior to initiating the slag foaming experiments, a quantity of approximately 5 g of ultra-low carbon (ULC) steel, characterized by a carbon content of 0.007 wt.%, was positioned at the base of an alumina crucible. This crucible featured inner dimensions measuring 63 mm in diameter, a wall thickness of 4 mm, and a height of 99 mm. Subsequently, roughly 101 g of slag powder was introduced into the crucible, and the filled alumina crucible was placed within a larger graphite crucible featuring inner dimensions of 70.3 mm in diameter, a wall thickness of 20 mm, and a height of 202 mm. The temperature was elevated to 1600 °C and held for 1 h to ensure complete melting. Following the foaming tests, liquid nitrogen was employed to rapidly cool and solidify the foam structure, after which it was allowed to naturally cool to ambient temperature [29–31].

A video was recorded throughout the experiments to reveal the slag foam formation for each carbon source applied.

To measure the height of the slag foam, the difference between the top surface of the foam and the unfoamed layer at the bottom of the crucible was calculated. This measurement was conducted at three different positions within the crucible, and the average of these values provided the mean height of the slag foam.

Based on the slag foam height, the volume of slag before (V_{slag}) and after foaming (V_{foam}) was calculated using Equations (1) and (2):

$$V_{slag} = \frac{d^2 \cdot \pi}{4} \cdot h_{slag}, \tag{1}$$

$$V_{foam} = \frac{d^2 \cdot \pi}{4} \cdot h_{foam}, \tag{2}$$

where d is the inner diameter of the crucible, cm; h_{slag} is slag height before foaming, cm; h_{foam} is slag height after foaming, cm.

The relative slag foaming volume ($\frac{\Delta V}{V_0}$) was calculated by Equation (3):

$$\frac{\Delta V}{V_0} = \frac{V_{foam} - V_{slag}}{V_{slag}}, \tag{3}$$

Thermogravimetric analysis (TGA)/differential thermogravimetry (DTG) was carried out by SDT Q600 V20.9 Build 20 (TA Instruments, New Castle, DE, USA) to analyze the air combustion process of carbon sources. The carbon source samples (30.0 $\pm$ 1.0 mg) were loaded into an alumina pan (Al_2O_3), and the air flow rate was set to 100.0 mL$\cdot$min^{-1} during test runs. In this experiment, the temperature was raised to 1100.0 °C from room temperature, with a heating rate of 5 °C/min under the air. The same experimental conditions were used for all of the tests with carbon sources. The change in the mass of the samples during heating was recorded to obtain the TG-DTG curves using the software OriginPro 2023b (version 10.0.5.157). The baseline for mass loss was obtained using an empty alumina pan.

The ignition and burnout temperatures of the samples were determined via the intersection method, according to [32]. The Coats–Redfern method was used to estimate the activation energy (E_a) on the combustion zone between the ignition and the burnout temperature [18,33,34].

FTIR spectroscopy analysis used an attenuated total reflectance (ATR) accessory with an Agilent Cary 630 FTIR spectrometer manufactured by Agilent Technologies, Santa Clara, CA, USA. For each sample, 32 scans were performed within the wavenumber range of 4000–650 cm^{-1}, employing a spectral resolution of 2 cm^{-1} and achieving a high wavenumber accuracy of 0.05 cm^{-1}. The software used for peaks analysis of the obtained spectra was OriginPro 2023b (version 10.0.5.157).

Afterward, using Bragg's Law, the aromatic planes of carbon crystallites (d_{002}) and the carbon crystallite width (L_a) were calculated according to [35,36] using OriginPro 2023b (version 10.0.5.157) based on X-ray diffraction (XRD) spectra of carbon source samples that were obtained using a Bruker AXS D8 advance diffractometer (Bruker Corporation, Billerica, MA, USA) with a lynxeye detector and a Cu X-ray tube with Cu Kα radiation.

3. Results and Discussion

3.1. Thermogravimetric Characters of Carbon Sources

The reactivity of the carbon source samples can be studied via the TGA, and the results for the determination of mass change rates due to combustion under air are reported. The weight of the samples were recorded using the TGA technique, and the TG–DTG curves were obtained for the coke and biocoke samples. According to the DTG profiles (Figure 1a–c), the three carbon source samples show curves with a single definite peak rate;

so, in this case, the combustion should be divided, generally, into three stages: (1) start of combustion and mass loss of coke/biocoke samples. (2) The second stage is a reaction with air oxygen. The coke/biocoke sample loses weight due to the combustion of fixed carbon, and the reaction rate is determined by the structural features of the carbon source. (3) In the final stage, after burning, the combustible components in coke/biocoke continue to decompose at a very low rate, producing ash as the final residue. The curves show a plateau for all three samples in the region of 750 °C and higher.

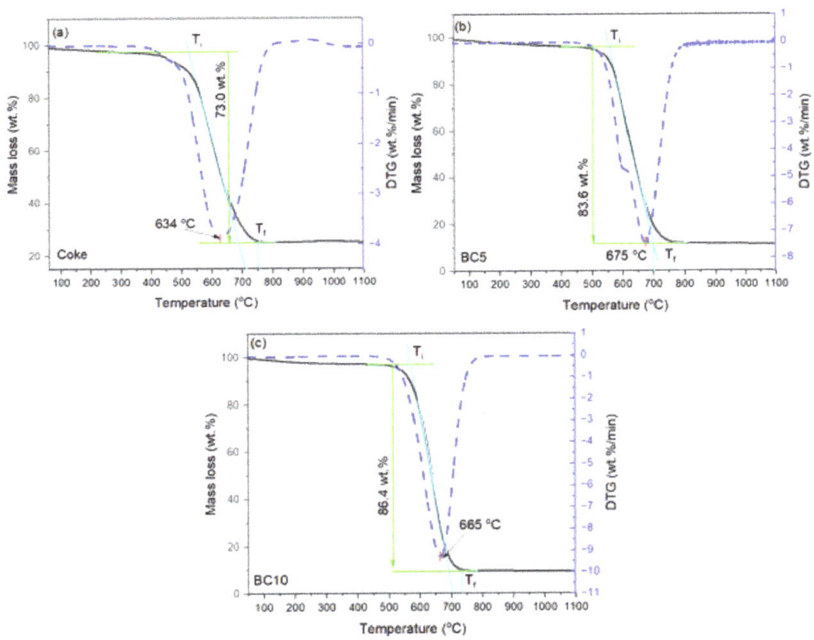

Figure 1. TG-DTG curves of carbon source combustion under air: (**a**) coke; (**b**) BC5; and (**c**) BC10. Assignments of characters are given within the text.

Notably, for all three samples, the initial stage of gradual weight loss due to moisture removal up to 100 °C and adsorbed gases up to 250 °C is practically unnoticeable. This is typically the case with coal or biomass/biochar; however, since coke or biocoke is a more heat-treated carbon material, this stage is not noticeable from the curves, considering low moisture and volatile matters, although it is evident from the proximate analysis in Table 2.

Regarding the mass loss curve, the main stage spans the region where the release and combustion of fixed carbon are responsible for the mass loss of coke and biocoke samples. Coke is characterized by the lowest weight loss and reaches 73.0 wt.% at 634 °C. The weight loss observed in biocoke samples is significant. When comparing the biocoke samples, the sample containing 5 wt.% wood pellets shows a slightly lower weight loss of 83.6 wt.% at 675 °C, whereas the biocoke sample with 10 wt.% wood pellets exhibits a higher weight loss of 86.4 wt.% at 665 °C.

A distinctive phenomenon is the significant difference in mass loss between the coke and the two biocoke samples. Additionally, there is a less noticeable difference in weight loss between biocoke samples. Considering the ash content for coke of 11.1 wt.%–db (Table 2), it can be concluded that this carbon source did not burnout completely during the TGA process. It can be explained by the fact that in the pan, the ash particles covering the surface of the carbon source particles block the pores present on the surface of the coke sample, increasing the extent of diffusion resistance realized. This significantly influences combustion reactivity and the degree of burnout. This observation was also noticed in [17].

In addition, since the coke did not burn out completely, this also explains the low peak temperature (T_p) compared to biocokes.

Moreover, the combustion characteristic indexes of the coke/biocoke can be obtained through the calculation and analysis of TG and DTG curves, which are mainly represented by ignition temperature (T_i) and peak temperature (T_p), which reveal the temperature of maximum mass loss, burnout temperature (T_f), and ΔT, which represents the difference between ignition temperature and burnout temperature [18]. The combustion characteristics of carbon source samples, the temperatures of coke, and two biocoke samples are shown in Table 4.

Table 4. Combustion characteristics of carbon source samples.

Combustion Characteristics	Coke	BC5	BC10
Ignition temperature T_i, °C	530	558	559
Peak temperature T_p, °C	634	675	665
Burnout temperature T_f, °C	758	701	735
ΔT, °C	228	143	176
Weight loss, wt.%	73.0	83.6	86.4

Higher burnout temperatures T_f are explained by the presence of the most stable components in the carbon material. Generally, T_i and T_f temperatures would be low if the reactivity of the carbon sources were high. However, the values of T_i and T_f on the TG curves turned out to be uncertain. The ignition temperature does not differ much between biocokes. The maximum burnout temperature for coke is 758 °C. The ΔT value of 228 °C may indicate a stronger coke structure and lower reactivity than biocoke samples. The ΔT value indicates that the biocoke samples' combustion duration was shorter than that of coke. However, the ignition temperature is lower for coke compared with the biocoke samples.

To explain this, it requires considering the microstructural aspects of the carbon sources. The interlayer spacing between aromatic planes of carbon crystallites (d_{002}) was calculated, obtaining a value of 0.36 nm for coke and 0.35 nm for the biocoke samples. Typically, a more minor d_{002} indicates a more ordered carbon structure and a higher degree of graphitization [37]. In this case, the similar d_{002} value across coke and the two biocoke samples suggests a comparable level of structural order. This similarity can be attributed to the relatively small proportion of wood pellet additives, which did not significantly influence the microstructural characteristics of the biocoke. Additionally, when analyzing biocokes, the appropriate distribution of the main material of coke and wood pellet particles in the sample is important since the indicators depend on the distribution of pyrolyzed wood pellets in the test sample. The comparable values of d_{002} prevent us from making decisive assessments regarding the impact of microstructure on explaining combustion characteristics. On the other hand, the L_a index indicates that coke has slightly better microstructural properties [38,39] compared to biocokes.

If we focus on the correlation between combustion characters and the effect of carbon source properties (Table 2), the volatile matter and fixed carbon values do not significantly lead to conclusive observations as these values are closely aligned. However, a general criterion suggests that the carbon source burns at lower temperatures when it is more hydrogenated [40]. The reason behind the earlier start of coke ignition compared to biocoke samples might be attributed to the higher hydrogen content in coke (0.30 wt.%−db) in contrast to biocoke samples with 0.27 wt.%−db and 0.22 wt.%−db for biocoke with 5 wt.% wood pellets and biocoke with 10 wt.% wood pellets, respectively.

Assessing the mass loss kinetics during the combustion of coke or biocokes is challenging due to the complex aromatic substances and the influence of multiple factors. An Arrhenius-type kinetic model was used to analyze the TG-DTG data. Based on DTG curves for all three carbon samples at a constant heating rate (5 °C/min), a single peak was observed in the range from ignition temperature to burnout temperature. This indicates that the combustion process is a single-step process. The activation energy results for the three carbon source samples are shown in Figure 2. E_a reflects the minimum energy that the reactant molecules need to reach the activated state during a chemical reaction. The higher activation energy value for coke, at 158.56 kJ/mol, suggests that more energy is required to initiate the reaction for coke compared to other carbon sources.

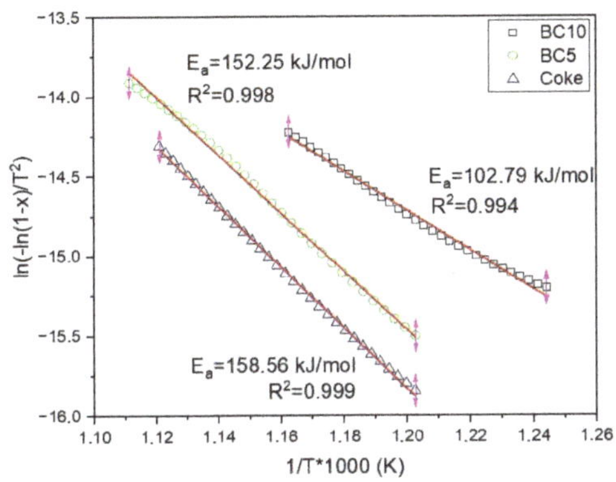

Figure 2. Arrhenius curves calculated for three carbon sources.

The activation energy values differed among the samples. The biocoke with 5 wt.% wood pellets exhibited a slightly lower activation energy of 152.25 kJ/mol, implying a noticeable but not significantly influential impact due to the addition of wood pellets. The lowest activation energy value was recorded in the biocoke sample containing 10 wt.% wood pellets at 102.79 kJ/mol. This substantial decrease in activation energy is noteworthy compared to both coke and biocoke with 5 wt.% wood pellets.

Although the activation energy values for coke and biocoke with 5 wt.% wood pellets were close to the values specified in [18] for a heating rate of 5 °C/min calculated by the Coats–Redfern method, they were marginally lower. This difference can be attributed to the fact that the samples in this study were laboratory-based. Adding 10 wt.% wood pellets to produce biocoke significantly contributed to decreased activation energy.

An FTIR analysis was conducted to confirm the occurrence of incomplete burnout in the TGA combustion of coke. In Figure 3, FTIR spectroscopy peaks and their corresponding assignments for carbon materials are depicted after conducting TGA. The O=C=O stretching vibrations, characteristic of carbon dioxide (CO_2), were identified at distinct wavenumbers for the different carbon sources: 2361 cm^{-1} for coke, 2360 cm^{-1} for biocoke with 10 wt.% wood pellets, and 2354 cm^{-1} for biocoke with 5 wt.% wood pellets.

The presence of oxygen-containing groups [33] is specific to coke, which is indicated by a C–O stretching peak (related to primary alcohol) at 1054 cm^{-1} and 1042 cm^{-1} for biocoke with 5 wt.% wood pellets. Meanwhile, biocoke with 10 wt.% wood pellets features a peak at 1132 cm^{-1}, signifying aliphatic ether functionality. Transmittance peaks at 900–700 cm^{-1} refer to the C–H bending vibrations of aromatic hydrocarbons, which sharply decrease for biocoke with 10 wt.% wood pellets, indicating the elimination of functional groups of aromatic compounds, resulting in a change in their relative –CH content. For biocoke with 10 wt.% wood pellets, the intensity of the peaks is very weak compared to coke or biocoke with 5 wt.% wood pellets, indicating that biocoke with 10 wt.% wood pellets has almost entirely reacted with air oxygen.

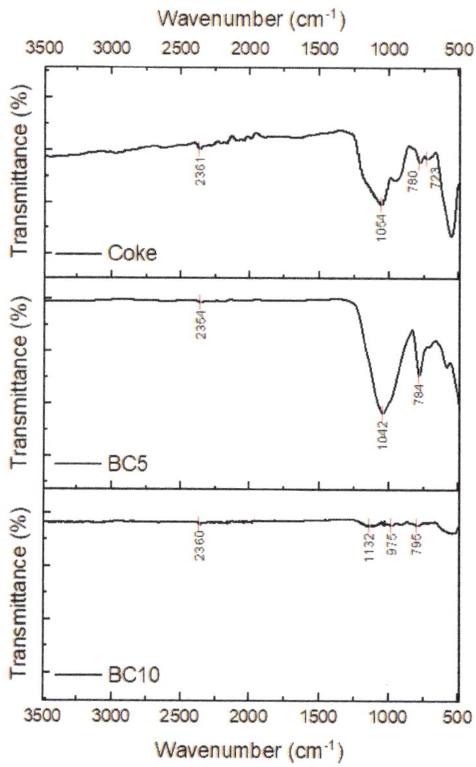

Figure 3. FTIR spectra of three carbon sources after combustion. Assignments of peaks are given within the text.

3.2. Formation of Slag Foam and Its Characteristics

It is worth recognizing that the specified test methodology of slag foaming has some peculiarity. Since the carbon material was supplied from above to the molten slag, taking into account the high temperature of 1600 °C, it was possible to observe the combustion process of some of the carbon source particles on the slag surface. With a more reactive carbon source, this process was generally more pronounced. Thus, when using coke, this was weakly noticed and more pronounced when using biocoke samples. Further, the carbon source particles floated on the molten slag surface, and slag foaming formations were initiated on the slag surface simultaneously. Because of this, quick mixing of the carbon source and molten slag was required each time. Nonetheless, slag foam formation is presented in Figure 4 for each carbon source applied. Slag foam formation is segmented to show this process from the initiation of foaming to its final stage. Notably, the duration of foaming varied among the three cases; however, the stages and evolution of slag foam exhibited minor differences. When coke was employed (Figure 4a), the foaming process showed the longest duration, extending to approximately 580 s. In contrast, the formation of equivalent stages during slag foaming with biocoke containing 5 wt.% wood pellets (Figure 4b) occurred over approximately 417 s. Subsequently, with biocoke incorporating 10 wt.% wood pellets, slag foaming was completed in about 185 s (Figure 4c). Segment 1 on Figure 4a–c depicts the initial state of the system before introducing the carbon source, representing the initial height of the molten slag in the crucible. The time gap between Segment 1 and Segment 2 is caused by the time of addition of the carbon source and the slight mixing of the carbon source and molten slag. Upon introducing the carbon source, the stage of active foaming began immediately, characterized by a noticeable increase in both the height and volume of the slag foam. Segments 2–3 in Figure 4a–c represent the active stage in slag foaming.

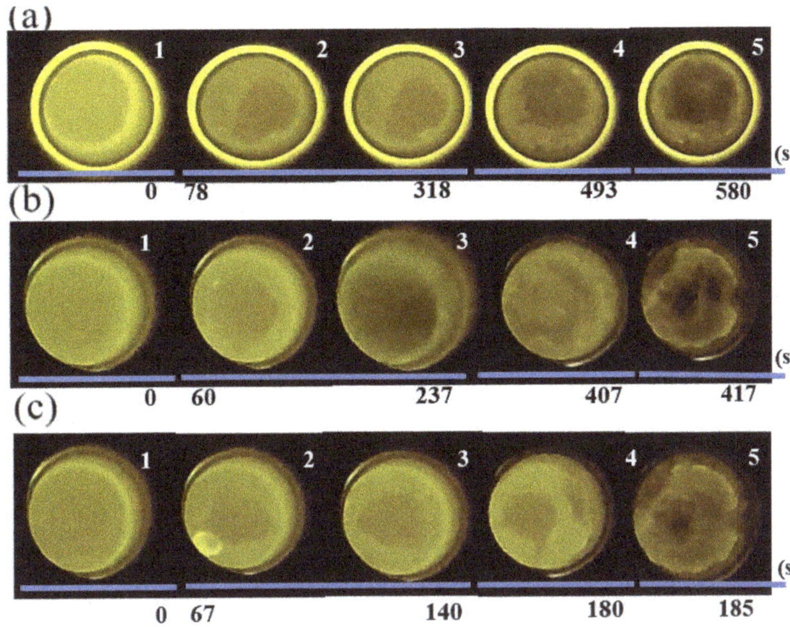

Figure 4. Slag foaming formation after adding carbon source: (**a**) coke; (**b**) BC5; (**c**) BC10.

A notable difference at this stage was that when using biocoke with 10 wt.% wood pellets, foaming was characterized by a slightly chaotic system, with the formation of large bubbles near the crucible walls, which led to some minor oscillation of the slag foam during its formation. Nevertheless, this minimal oscillation did not substantially influence the overall formation of slag foam. Conversely, such oscillations were less pronounced when employing the other two carbon sources. Subsequently, the foaming behavior entered a more stable stage (Segment 4), characterized by a gradual change in the height and volume of the slag foam.

The last segment, Segment 5 of Figure 4a–c, represents the final stage of the foaming process. This phase is characterized by the absence of changes in the height and volume of the slag foam aligned with the almost completely reacted carbon source.

In all three cases, stable foaming was observed; upon adding a carbon source, there was a noticeable increase in both volume and height without significant fluctuations or uncontrolled slag foaming. Coke demonstrated generally average slag foaming behavior (Table 5). However, when 5 wt.% wood pellets were added to the biocoke blends, the foaming behavior notably improved. Conversely, using biocoke with 10 wt.% wood pellets deteriorated slag foaming behavior compared to the previous carbon sources. Specifically, the relative foaming volume ($\Delta V/V_0$) was higher for biocoke with 5 wt.% wood pellets compared to coke, while for biocoke with 10 wt.% wood pellets, it was slightly worse than coke. Typically, substituting traditional carbon sources with biocoke may potentially reduce and limit foaming time, which largely depends on the reactivity of the carbon source [41]. Likewise, it is worth adding that in the case of laboratory tests, the height and volume of foaming are also limited by the dimensions of the crucible.

Table 5. Slag foaming characters based on a single test for each carbon source.

Slag Characters	Molten Slag		
Initial slag height mean h_{slag}, cm	0.9		
Initial slag volume mean V_{slag}, cm^3	28.04		
Slag foaming characters	**SF (Coke)**	**SF (BC5)**	**SF (BC10)**
Slag foam height h_{foam}, cm	7.8	8.6	7.2
Time of slag foaming, s	~580	~417	~185
Slag foam volume V_{foam}, cm^3	243.02	267.94	224.33
Relative foaming volume, $\Delta V/V_0$	7.67	8.56	7.00

Figure 5 depicts the variations in slag height subsequent to the addition of a carbon source. It shows significant changes in height compared to the initial height of the molten slag. The highest height was observed when utilizing biocoke with 5 wt.% wood pellets, reaching 8.6 cm. In contrast, using coke resulted in a foaming height of 7.8 cm, with a foaming time of approximately 580 s. The use of biocoke notably reduced the foaming time, particularly evident when employing biocoke with 10 wt.% wood pellets. This variant of biocoke exhibited the smallest foaming height, likely due to insufficient reaction time, leading to a decrease in the observed height.

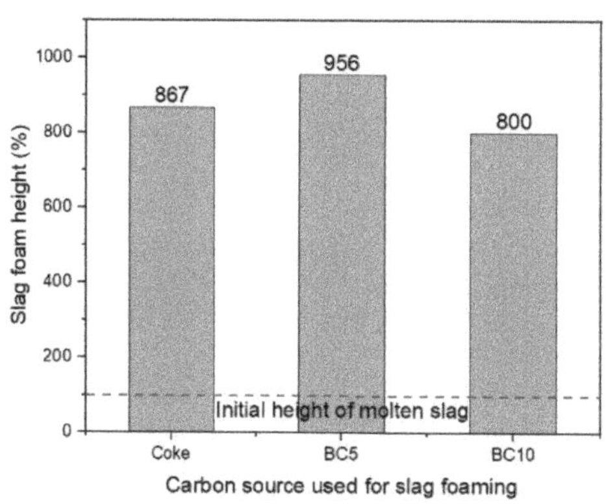

Figure 5. Changes in slag foam height depending on the carbon source applied.

It is known that the addition of 5 wt.% wood pellets and more in biocoke causes increased reactivity compared to standard metallurgical coke [42,43]. The changes in the physicochemical and microstructural characteristics of coke are a result of the introduction of biomass pellet additives.

When using biocoke with 5 or 10 wt.% wood pellets, it can be assumed that the foaming time was reduced due to an accelerated reaction rate. However, it is essential to admit that numerous factors influence slag foaming, and the reactivity of carbon sources is only one factor. Given the constant slag composition and test conditions, the properties of the carbon sources and their wettability could play a role in influencing foaming behavior. The ash content, volatile matters, and fixed carbon across all three carbon sources remained the same and may not be significant in this context. While the d_{002} values are nearly identical, microstructural properties cannot entirely explain the effect on slag foaming.

Additionally, although the L_a values for coke are notably higher, the values for both biocoke samples are similar. Hence, microstructural properties might not be sufficient to explain the influence on slag foaming.

As for the interaction behavior of the carbon/slag system, it is one of the most important factors and could significantly influence the reactions occurring. Coke is characterized by better interaction with slag compared to biochar, as was reported in [8,44], and it has been reported that the interaction between slag and biochar has the lowest interaction at 1600 °C [8]. Based on this, it can be assumed that the partial replacement of coal with wood pellets could hardly improve the interaction behavior of the carbon/slag system and therefore could hardly influence and explain the phenomenon of enhanced foaming when using biocoke with 5 wt.% wood pellets.

In this study, the test conditions and slag composition were equal, and the only variable factor was the carbon source. If we consider the activation energy, then biocoke with 5 wt.% wood pellets requires slightly less energy to react, as this is a more reactive carbon source compared to coke. A decrease in the time of slag foaming also evidences this. When using biocoke with 10 wt.% wood pellets, the foaming time was decreased significantly by ~185 s, and the lowest activation energy value indicates that this carbon source is likely to react more easily and quickly. It can be assumed that in this case, this became a limiting factor for the formation of slag foaming, which led to lower values of foaming characters in comparison with biocoke with 5 wt.% wood pellets. Indeed, apart from the mentioned factors that can influence slag foaming behavior, other factors should also be taken into account, as presented earlier in [45–47], such as the reactivity of carbon sources with FeO, the reactivity of carbon sources with CO_2, the reactivity of carbon sources in molten slag containing FeO (liquid) with CO_2, and, finally, CO/CO_2 concentration in gas.

In summary, it should be recognized that the impact of carbon sources on slag foaming behavior is versatile. The confirmed practical potential of utilizing biocoke as a carbon source for slag foaming through laboratory-scale tests emphasizes the importance of understanding the physicochemical, microstructural, and thermochemical properties, as well as the relationships between carbon source properties when assessing their applicability for this specific purpose.

4. Conclusions

This study investigated the impact of the thermochemical properties of two biocoke variants, differentiated by 5 wt.% and 10 wt.% wood pellet additions as carbon sources for facilitating foamy slag formations. Conducted at 1600 °C under consistent test conditions, the comparison was made against a laboratory-scale coke sample used as a reference. Based on the research results, the following conclusions were made:

- The reactivity of the carbon source samples through combustion under air revealed the difference in mass loss between carbon sources in increasing order of coke < biocoke with 5 wt.% wood pellets < biocoke with 10 wt.% wood pellets. Likewise, differences were revealed in the activation energies between these sources, where coke E_a was 158.56 kJ/mol, for biocoke with 5 wt.% wood pellets, it was 152.25 kJ/mol, and for biocoke with 10 wt.% wood pellets, it was 102.79 kJ/mol.
- According to FTIR analysis, for biocoke with 10 wt.% wood pellets, the intensity of the peaks is very weak compared to coke or biocoke with 5 wt.% wood pellets, indicating that biocoke with 10 wt.% wood pellets has almost entirely reacted with air oxygen.
- The applicability of biocoke for facilitating foaming slag was confirmed, marked by stable foaming and increased volume and height, without an uncontrollable change in the slag foaming characters. Comparatively, the relative foaming volume, $\Delta V/V_0$, was improved for biocoke with 5 wt.% wood pellets in contrast to coke. However, the utilization of biocoke with 10 wt.% wood pellets exhibited marginally less foaming compared to coke. The introduction of 5 wt.% wood pellets increased the reaction rate,

reducing slag foaming time. Conversely, adding 10 wt.% wood pellets significantly decreased foaming time, limiting this process.

This study emphasizes that the influence of carbon sources is a complex consideration, with various factors affecting the overall behavior. Characters such as ash content, volatile matters, fixed carbon value, and microstructural features of carbon sources are important and should be considered. However, based on the results, the reactivity of carbon sources as a complex of qualitative characteristics is of great importance. Therefore, the thermochemical properties of carbon sources should be considered when evaluating the applicability of unconventional carbon sources for slag foaming.

Author Contributions: Conceptualization, L.K. and J.S.; methodology, L.K. and J.S.; investigation, L.K., A.K. and A.H.; writing—original draft preparation, L.K.; writing—review and editing, J.S.; supervision, J.S. All authors have read and agreed to the published version of the manuscript.

Funding: This research received no external funding.

Data Availability Statement: Relevant data are contained within the article.

Acknowledgments: This work was supported by the scholarship program "Scholarship of the Scholarship Foundation of the Republic of Austria, Postdocs", [MPC-2022-02241], financed by the Federal Ministry of Education, Science and Research of Austria, which is gratefully acknowledged. We would like to acknowledge the staff of the Chair of Ferrous Metallurgy, who supported us throughout this research.

Conflicts of Interest: Author Lina Kieush was employed by the company K1-MET GmbH. The remaining authors declare that the research was conducted in the absence of any commercial or financial relationships that could be construed as a potential conflict of interest.

References

1. Axelson, M.; Oberthür, S.; Nilsson, L.J. Emission Reduction Strategies in the EU Steel Industry: Implications for Business Model Innovation. *J. Ind. Ecol.* **2021**, *25*, 390–402. [CrossRef]
2. Hara, S.; Ogino, K. Slag-Foaming Phenomenon in Pyrometallurgical Processes. *ISIJ Int.* **1992**, *32*, 81–86. [CrossRef]
3. Luz, A.P.; Avila, T.A.; Bonadia, P.; Pandolfelli, V.C. Slag Foaming: Fundamentals, Experimental Evaluation and Application in the Steelmaking Industry. *Refract. Worldforum* **2011**, *3*, 91–98.
4. Agnihotri, A.; Singh, P.K.; Singh, D.; Gupta, M. Foamy Slag Practice to Enhance the Energy Efficiency of Electric Arc Furnace: An Industrial Scale Validation. *Mater. Today Proc.* **2021**, *46*, 1537–1542. [CrossRef]
5. Morales, R.D.; Rubén, L.G.; López, F.; Camacho, J.; Romero, J.A. The Slag Foaming Practice in EAF and Its Influence on the Steelmaking Shop Productivity. *ISIJ Int.* **1995**, *35*, 1054–1062. [CrossRef]
6. Matsuura, H.; Fruehan, R.J. Slag Foaming in an Electric Arc Furnace. *ISIJ Int.* **2009**, *49*, 1530–1535. [CrossRef]
7. Vieira, D.; Almeida, R.A.M.D.; Bielefeldt, W.V.; Vilela, A.C.F. Slag Evaluation to Reduce Energy Consumption and EAF Electrical Instability. *Mat. Res.* **2016**, *19*, 1127–1131. [CrossRef]
8. Huang, X.-A.; Ng, K.W.; Giroux, L.; Duchesne, M. Carbonaceous Material Properties and Their Interactions with Slag During Electric Arc Furnace Steelmaking. *Met. Mater. Trans. B* **2019**, *50*, 1387–1398. [CrossRef]
9. Kieush, L.; Koveria, A.; Boyko, M.; Yaholnyk, M.; Hrubiak, A.; Molchanov, L.; Moklyak, V. Influence of Biocoke on Iron Ore Sintering Performance and Strength Properties of Sinter. *Min. Min. Depos.* **2022**, *16*, 55–63. [CrossRef]
10. Solar, J.; Hernandez, A.; Lopez-Urionabarrenechea, A.; De Marco, I.; Adrados, A.; Caballero, B.M.; Gastelu, N. From Woody Biomass Waste to Biocoke: Influence of the Proportion of Different Tree Components. *Eur. J. Wood Prod.* **2017**, *75*, 485–497. [CrossRef]
11. Koveria, A.; Kieush, L.; Svietkina, O.; Perkov, Y. Metallurgical Coke Production with Biomass Additives. Part 1. A Review of Existing Practices. *Can. Metall. Q.* **2020**, *59*, 417–429. [CrossRef]
12. Montiano, M.G.; Díaz-Faes, E.; Barriocanal, C.; Alvarez, R. Influence of Biomass on Metallurgical Coke Quality. *Fuel* **2014**, *116*, 175–182. [CrossRef]
13. Hu, M.; Chen, Z.; Wang, S.; Guo, D.; Ma, C.; Zhou, Y.; Chen, J.; Laghari, M.; Fazal, S.; Xiao, B.; et al. Thermogravimetric Kinetics of Lignocellulosic Biomass Slow Pyrolysis Using Distributed Activation Energy Model, Fraser–Suzuki Deconvolution, and Iso-Conversional Method. *Energy Convers. Manag.* **2016**, *118*, 1–11. [CrossRef]
14. Chen, X.; Liu, L.; Zhang, L.; Zhao, Y.; Qiu, P. Pyrolysis Characteristics and Kinetics of Coal–Biomass Blends during Co-Pyrolysis. *Energy Fuels* **2019**, *33*, 1267–1278. [CrossRef]
15. Chen, W.-H.; Wu, J.-S. An Evaluation on Rice Husks and Pulverized Coal Blends Using a Drop Tube Furnace and a Thermogravimetric Analyzer for Application to a Blast Furnace. *Energy* **2009**, *34*, 1458–1466. [CrossRef]

16. Xu, J.; Zuo, H.; Wang, G.; Zhang, J.; Guo, K.; Liang, W. Gasification Mechanism and Kinetics Analysis of Coke Using Distributed Activation Energy Model (DAEM). *Appl. Therm. Eng.* **2019**, *152*, 605–614. [CrossRef]
17. Yang, T.; Wei, B.; Wang, S.; Liu, W.; Chen, L.; Wang, J.; Abudureheman, M.; Ma, J.; Wang, F. Effect of Particle Size on the Kinetics of Pure Oxygen Combustion of Coke. *Thermochim. Acta* **2023**, *719*, 179405. [CrossRef]
18. Qin, Y.; Ling, Q.; He, W.; Hu, J.; Li, X. Metallurgical Coke Combustion with Different Reactivity under Nonisothermal Conditions: A Kinetic Study. *Materials* **2022**, *15*, 987. [CrossRef]
19. Fidalgo, B.; Berrueco, C.; Millan, M. Chars from Agricultural Wastes as Greener Fuels for Electric Arc Furnaces. *J. Anal. Appl. Pyrolysis* **2015**, *113*, 274–280. [CrossRef]
20. Cardarelli, A.; De Santis, M.; Cirilli, F.; Barbanera, M. Computational Fluid Dynamics Analysis of Biochar Combustion in a Simulated Ironmaking Electric Arc Furnace. *Fuel* **2022**, *328*, 125267. [CrossRef]
21. Wei, R.; Zheng, X.; Zhu, Y.; Feng, S.; Long, H.; Xu, C.C. Hydrothermal Bio-Char as a Foaming Agent for Electric Arc Furnace Steelmaking: Performance and Mechanism. *Appl. Energy* **2024**, *353*, 122084. [CrossRef]
22. Son, K.; Lee, J.; Hwang, H.; Jeon, W.; Yang, H.; Sohn, I.; Kim, Y.; Um, H. Slag Foaming Estimation in the Electric Arc Furnace Using Machine Learning Based Long Short-Term Memory Networks. *J. Mater. Res. Technol.* **2021**, *12*, 555–568. [CrossRef]
23. Yunos, N.F.M.; Zaharia, M.; Ismail, A.N.; Idris, M.A. Transforming Waste Materials as Resources for EAF Steelmaking. *Int. J. Mater. Eng.* **2014**, *4*, 167–170.
24. Zhang, Y.; Fruehan, R.J. Effect of the Bubble Size and Chemical Reactions on Slag Foaming. *Met. Mater Trans. B* **1995**, *26*, 803–812. [CrossRef]
25. Luz, A.P.; Tomba Martinez, A.G.; López, F.; Bonadia, P.; Pandolfelli, V.C. Slag Foaming Practice in the Steelmaking Process. *Ceram. Int.* **2018**, *44*, 8727–8741. [CrossRef]
26. Menad, N.-E.; Kana, N.; Seron, A.; Kanari, N. New EAF Slag Characterization Methodology for Strategic Metal Recovery. *Materials* **2021**, *14*, 1513. [CrossRef]
27. *ASTM D3172-13*; Standard Practice for Proximate Analysis of Coal and Coke 2013. ASTM International: West Conshohocken, PA, USA, 2013.
28. Gudenau, H.W.; Senk, D.; Fukada, K.; Babich, A.; Froehling, C. Coke Behaviour in the Lower Part of BF with High Injection Rate. In *International BF Lower Zone Symposium*; Illawarra Branch: Wollongong, Australia, 2002.
29. Kieush, L.; Koveria, A.; Schenk, J.; Rysbekov, K.; Lozynskyi, V.; Zheng, H.; Matayev, A. Investigation into the Effect of Multi-Component Coal Blends on Properties of Metallurgical Coke via Petrographic Analysis under Industrial Conditions. *Sustainability* **2022**, *14*, 9947. [CrossRef]
30. Kieush, L.; Schenk, J. Investigation of the Impact of Biochar Application on Foaming Slags with Varied Compositions in Electric Arc Furnace-Based Steel Production. *Energies* **2023**, *16*, 6325. [CrossRef]
31. Kieush, L.; Schenk, J.; Koveria, A.; Hrubiak, A.; Hopfinger, H.; Zheng, H. Evaluation of Slag Foaming Behavior Using Renewable Carbon Sources in Electric Arc Furnace-Based Steel Production. *Energies* **2023**, *16*, 4673. [CrossRef]
32. Lu, J.-J.; Chen, W.-H. Investigation on the Ignition and Burnout Temperatures of Bamboo and Sugarcane Bagasse by Thermogravimetric Analysis. *Appl. Energy* **2015**, *160*, 49–57. [CrossRef]
33. Coats, A.W.; Redfern, J. Kinetic Parameters from Thermogravimetric Data. *Nature* **1964**, *201*, 68. [CrossRef]
34. Tang, L.; Xiao, J.; Mao, Q.; Zhang, Z.; Yao, Z.; Zhu, X.; Ye, S.; Zhong, Q. Thermogravimetric Analysis of the Combustion Characteristics and Combustion Kinetics of Coals Subjected to Different Chemical Demineralization Processes. *ACS Omega* **2022**, *7*, 13998–14008. [CrossRef] [PubMed]
35. Okolo, G.N.; Neomagus, H.W.J.P.; Everson, R.C.; Roberts, M.J.; Bunt, J.R.; Sakurovs, R.; Mathews, J.P. Chemical–Structural Properties of South African Bituminous Coals: Insights from Wide Angle XRD–Carbon Fraction Analysis, ATR–FTIR, Solid State 13 C NMR, and HRTEM Techniques. *Fuel* **2015**, *158*, 779–792. [CrossRef]
36. Warren, B.E. X-Ray Diffraction in Random Layer Lattices. *Phys. Rev.* **1941**, *59*, 693–698. [CrossRef]
37. Wang, J.; Wang, W.; Chen, X.; Bao, J.; Duan, L.; Xu, R.; Zheng, H. Investigation on the Evolution of Structure and Strength of Iron Coke during Carbonization: Carbon Structure, Pore Structure, and Carbonization Mechanism. *Powder Technol.* **2024**, *431*, 119059. [CrossRef]
38. Zheng, H.; Xu, R.; Zhang, J.; Daghagheleh, O.; Schenk, J.; Li, C.; Wang, W. A Comprehensive Review of Characterization Methods for Metallurgical Coke Structures. *Materials* **2021**, *15*, 174. [CrossRef] [PubMed]
39. Fu, X.; Pang, Q.; Yang, X.; Zhan, W.; He, Z. Effect of High Temperature on Macroscopic Properties and Microstructure of Metallurgical Coke. *Fuel* **2024**, *356*, 129543. [CrossRef]
40. Ochoa, A.; Ibarra, Á.; Bilbao, J.; Arandes, J.M.; Castaño, P. Assessment of Thermogravimetric Methods for Calculating Coke Combustion-Regeneration Kinetics of Deactivated Catalyst. *Chem. Eng. Sci.* **2017**, *171*, 459–470. [CrossRef]
41. DiGiovanni, C.; Li, D.; Ng, K.W.; Huang, X. Ranking of Injection Biochar for Slag Foaming Applications in Steelmaking. *Metals* **2023**, *13*, 1003. [CrossRef]
42. Ng, K.W.; MacPhee, J.A.; Giroux, L.; Todoschuk, T. Reactivity of Bio-Coke with CO_2. *Fuel Process. Technol.* **2011**, *92*, 801–804. [CrossRef]
43. Suopajärvi, H.; Dahl, E.; Kemppainen, A.; Gornostayev, S.; Koskela, A.; Fabritius, T. Effect of Charcoal and Kraft-Lignin Addition on Coke Compression Strength and Reactivity. *Energies* **2017**, *10*, 1850. [CrossRef]

44. Ogawa, Y.; Katayama, H.; Hirata, H.; Tokumitsu, N.; Yamauchi, M. Slag Foaming in Smelting Reduction and Its Control with Carbonaceous Materials. *ISIJ Int.* **1992**, *32*, 87–94. [CrossRef]
45. Oh, J.S.; Lee, J. Composition-Dependent Reactive Wetting of Molten Slag on Coke Substrates. *J. Mater. Sci.* **2016**, *51*, 1813–1819. [CrossRef]
46. Migas, P.; Karbowniczek, M. Interactions between Liquid Slag and Graphite During the Reduction of Metallic Oxides. *Arch. Metall. Mater.* **2010**, *55*, 1147–1157. [CrossRef]
47. Sahajwalla, V.; Khanna, R.; Rahman, M.; Saha-Chaudhury, N.; Knights, D.; O'Kane, P.; Association for Iron & Steel Technology. Recycling of Waste Plastics for Slag Foaming in EAF Steelmaking. In Proceedings of the AISTech—Iron and Steel Technology Conference Proceedings, Cleveland, OH, USA, 1–4 May 2006; Volume 2.

metals

MDPI

Article

The Behavior of Direct Reduced Iron in the Electric Arc Furnace Hotspot

Andreas Pfeiffer [1,*], Daniel Ernst [2], Heng Zheng [1], Gerald Wimmer [3] and Johannes Schenk [1]

1 Chair of Ferrous Metallurgy, Montanuniversität Leoben, Franz-Josef-Straße 18, 8700 Leoben, Austria; heng.zheng@unileoben.ac.at (H.Z.); johannes.schenk@unileoben.ac.at (J.S.)
2 K1-MET GmbH, Stahlstraße 14, 4020 Linz, Austria; daniel.ernst@k1-met.com
3 Primetals Technologies Austria GmbH, Turmstraße 44, 4031 Linz, Austria; gerald.wimmer@primetals.com
* Correspondence: andreas.pfeiffer@stud.unileoben.ac.at

Abstract: Hydrogen-based direct reduction is a promising technology for CO_2 lean steelmaking. The electric arc furnace is the most relevant aggregate for processing direct reduced iron (DRI). As DRI is usually added into the arc, the behavior in this area is of great interest. A laboratory-scale hydrogen plasma smelting reduction (HPSR) reactor was used to analyze that under inert conditions. Four cases were compared: carbon-free and carbon-containing DRI from DR-grade pellets as well as fines from a fluidized bed reactor were melted batch-wise. A slag layer's influence was investigated using DRI from the BF-grade pellets and the continuous addition of slag-forming oxides. While carbon-free materials show a porous structure with gangue entrapments, the carburized DRI forms a dense regulus with the oxides collected on top. The test with slag-forming oxides demonstrates the mixing effect of the arc's electromagnetic forces. The cross-section shows a steel melt framed by a slag layer. These experiments match the past work in that carburized DRI is preferable, and material feed to the hotspot is critical for the EAF operation.

Keywords: direct reduced iron; electric arc furnace; sponge iron; hydrogen DRI; slag formation

1. Introduction

Electric arc furnaces (EAF) produced more than 500 million t of crude steel in 2021 [1]. Although scrap is the primary iron carrier, the EAF is also a suitable aggregate for processing direct reduced iron (DRI) or hot briquetted iron (HBI). This combination of direct reduction reactor (DR) and EAF can be considered the major ore-based steelmaking strategy in natural gas (NG)-rich countries such as Mexico [2], Saudi Arabia [3,4], and United Arabic Emirates [5]. Nowadays, the 119 Mt DRI production in 2021 [6] is just a minor share of the total ferrous feed for global steel production. However, due to the ambitious climate targets of the European Union [7], this technology may gain greater importance in the future [8]. Therefore, the processing behavior of sponge iron is also of significant interest.

Many authors investigated the DRI melting behavior in the past. From the perspective of a single pellet, one must distinguish between the dissolution in liquid steel, hot metal, and slags. González et al. [9] mathematically describe the initial formation of a solid superficial layer on the pellet in contact with liquid iron. The layer increases the particle diameter from 12 to 18 mm, which starts to soften after approximately 6 s. Depending on the considered EAF arc length, the calculated diameter reaches 0 mm after 13 to 17 s. The melting rate critically depends on the initial pellet diameter. Pineda-Martínez et al. [10] report similar results in contact with non-reactive slags. However, due to the lower heat transfer, the melting times are dramatically higher compared to the liquid steel case. Besides the particle size, bath stirring is a decisive factor, as forced convection critically increases heat transport. Ramirez-Argaez et al. [11] further use CFD modeling to study the multiphase slag–steel–DRI system, with similar results. Pfeiffer et al. [12] confirm this trend experimentally. They submerge single samples of DRI and HBI into a liquid pool and quench the specimen after predefined times. Carbon-free DRI initially acquires solid layers from steel as well as slags. However, the higher the DRI carbon content, the faster the pellet liquefies in steel. Nevertheless, even highly carburized sponge iron remains solid in slag after three seconds of dipping. The authors explain this difference analytically using the Prandtl number, which is increased by a factor of approximately 1000 in the slag case. Samples dipped into saturated hot metal melt much faster, even if the melt temperature is below the iron liquidus temperature. Carbon diffusion from the liquid to the solid phase explains that. Penz et al. [13–15] investigate this phenomenon for scrap melting in basic oxygen furnace (BOF) conditions. After forming the initial solid shell, carbon diffuses as long as the liquidus temperature falls below the melting point. Once that happens, melting progresses.

DRI is usually continuously charged through the fifth hole in the EAF cover. To gain consistent flat bath operation, DRI-based EAF are practiced with up to 30% hot heel [16,17]. This operation is characterized by a short and stable arc mode and a smoothly increasing bath level. The ideal charging point is located in the hot spot, in the center of the three AC-EAF electrodes. Depending on the DRI temperature and the melting capacity, charging rates are limited to avoid the formation of so-called "icebergs" or "ferrobergs". Typical values range from 34 kg/(min·MW) [16,18,19] for cold DRI up to 55 kg/(min·MW) [20] for hot DRI. Consequently, hot DRI feeding is beneficial not only in terms of energy consumption, but also in terms of the charging rate. Therefore, lower tap-to-tap times can be expected.

In recent years, a lot of research has been conducted with a focus on electric steelmaking. The most important topics in this case are slag operation [17,21–24], process modelling [25,26], and application of alternative carbon sources [27–29]. These topics are of decisive importance when it comes to optimizing the EAF process for CO_2-neutral steelmaking in the future. Slag operation defines the process sequence in multiple aspects. On the one hand, slag foaming covers the electric arc and avoids radiation losses. Furthermore, it defines the yield of metallic iron since it is high in (FeO), the partition of harmful elements such as phosphorus, and the refractory lifetime. The use of biomass-based carbon carriers provides the chance to almost fully reduce fossil-based carbon emissions, as the electric arc furnace relies on a minimum amount of carbon for slag foaming operation. Lastly, process simulation and modelling enhance process efficiency. The better one's ability to simulate the EAF and its mass and energy balances, the more energy and input materials, such as slag forming oxides, can be saved.

However, there is still little data available in two respects. First is the application of DRI based on hydrogen. This kind of sponge iron is free of carbon. Therefore, its liquidus temperature is significantly higher than that of the conventional one. This difference may lead to a different behavior during the melting step in the EAF. Second is the direct observation of the feed material in the electric arc; it would be of great interest to learn more about its behavior. This study aims to generate deeper understanding in both mentioned respects and analyzes the behavior of DRI in an electric arc. Next to hydrogen-based carbon-free material, carburized samples and a continuous slag feed are tested. The observations provide more profound knowledge about the melting procedure and can be used to explain the optimal DRI charging strategy and deliver an idea about the final crude steel quality.

2. Materials and Methods

Table 1 lists the applied iron ores and their composition. An industry partner provided the samples and the chemical compositions. DR-grade pellets were used in tests 0, 1, and 2; a similar fine ore to test DRI fines from fluidized bed reduction was used in Test 3. To make the result of the slag Test 4 clearer, a BF-grade pellet with a higher gangue content was tested.

Table 1. Ore compositions in wt.-%.

	[2] Fe_{tot}	Fe^{2+}	SiO_2	Al_2O_3	CaO	MgO	TiO_2	MnO	Size/mm	Test No.
DR-Pellet	67.5	0.19	1.94	0.32	0.87	0.35	0.06	0.21	10–12.5	0, 1, 2
BF-Pellet	64.8	0.42	5.79	0.47	0.48	0.60	0.06	0.04	12.5–16	4
Fine-raw	66.8	6.19							D50 = 0.1	-
Fine-oxi [1]	66.4	1.87							D50 = 0.1	3

[1] Composition of oxidized fine ore was determined by the mass balance; [2] Fe_{tot}—Total iron content.

The pellet samples were reduced in a vertical reduction furnace (VRF) using a 75 mm diameter retort, precisely described in [30,31]. This is a standard aggregate used, for example, to perform standard tests such as ISO11258 [32]. The retort is fixed on a scale, providing the possibility to monitor the weight loss during the test. The reduction gas is preheated in the furnace chamber; it enters the retort from the bottom and exits through the top into the off-gas duct. The thermocouple measures the temperature in the center of the pellet load. Approximately 500 g pellets were charged. After heating up to 900 °C under nitrogen purging, 25 NL/min pure H_2 was applied. Once the mass loss reached the demanded metallization degree (MD), the test was stopped, and the material was cooled under a nitrogen atmosphere. Three pellet samples were prepared: two from the DR-pellets were reduced to 90% and 94% MD; one from the BF-pellets was reduced to 90% MD. In the same retort, the highly reduced DR-grade DRI was carburized afterward under 8 NL/min CH_4 at 800 °C. According to the mass balance, it contains 2.1% carbon; due to the high initial reduction, this is considered a reasonable presumption.

The fine ore was preoxidized in a heat treatment furnace [33] for 8 h at 700 °C. The prereduction step is essential as hematitic ore is beneficial in terms of reducibility and fluidization behavior. The sample was reduced in a 68 mm fluidized bed reactor. This reactor uses the same furnace as the 75 mm retort; its properties and dimensions are described in detail by Spreitzer and Schenk in [34]. Reduction was performed using 6 NL/min N_2 and 15 NL/min H_2 without application of additional pressure in the reduction chamber. A grid with holes of 0.4 mm in diameter was used, which is optimal for such fine-grained iron ores. The final MD is approximately 89.4%, which is determined by mass balance calculation.

The melting tests were performed in the hydrogen plasma smelting reduction (HPSR) reactor at the chair of ferrous metallurgy, Montanuniversität Leoben, Austria, whose layout is visualized in Figure 1. This aggregate, extensively described in [35,36], is usually applied for direct smelting reduction tests. The method itself is well established. Regarding reduction in atmosphere with hydrogen, numerous research topics were investigated in the past. Optimizations were conducted in case of arc stability [37], electrode shapes [38] and ore prereduction degree [39]. If operated in an inert atmosphere as in this test series, it can be considered a DC laboratory-scale electric arc furnace with a graphite electrode. The electrode had a diameter of 26 mm, a 10 mm tip, and a 5 mm axial opening through which the nitrogen was supplied.

The samples were charged batchwise in a steel crucible whose exact dimensions are summarized by Ernst et al. [39]; all experiments were performed with 2 NL/min nitrogen purging. Table 2 summarizes the performed tests. Test 0 acts as a method test and is not further evaluated. In Test 4, 14.0 g CaO and 3.1 g MgO were continuously fed through a hollow electrode. Technically pure oxides with 95 and 99% purity from Carl Roth GmbH + Co. KG, Karlsruhe, Germany were used. Initially, the transformer current was set to 100 A. During the test, it was decreased stepwise, depending on the transformer temperature and voltage. For that, a silicon-controlled rectifier (SCR) was applied. In the final approx. 2.5 min of Test 4, the power input was maximized to intensify the slag–metal stirring.

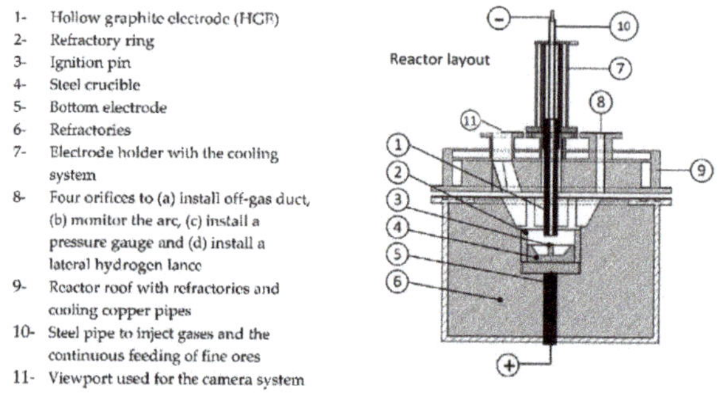

Figure 1. Schematic HPSR furnace layout reproduced under terms of the CC-BY license [37]. Copyright 2020, the authors, published by MDPI.

Table 2. List of tests.

Test	Gas	t/min	Sample	Mass/g	Comment
0		15	1	49.7	Pretest
1		15	1	101.6	0%C DR-grade-DRI
2	2 NL/min N_2	15	2	101.7	2.1%C DR-grade-DRI
3		15	3	100.4	0%C Fine-DRI
4		23	4	100.1	0%C BF-grade-DRI + Continuous feed of slag-forming oxides

The experiments were evaluated in multiple ways. The HPSR reactor had an AXIS-Q1775 camera system (Axis Communications AB, Lund, Sweden) and a GAM 200 mass spectrometer (Pfeiffer Vacuum Technologies AG, Aßlar, Germany), which provide the possibility to monitor the electric arc as well as the off-gas. Further, voltage and current values were recorded and documented.

The crucibles were visually checked using a Sony Alpha 6000 DSLM camera (Sony Group Corporation, Tokyo, Japan) equipped with a Sigma Contemporary 30 mm lens (Sigma Corporation, Kawasaki, Japan). Afterward, the crucibles were embedded, cut, and the cross-sections were metallographically prepared. Figure 2 details the cutting scheme. Etching was performed using a 1% nital solution to highlight the transition between the crucible and the sample. We adjusted the time iteratively between 6 and 10 s.

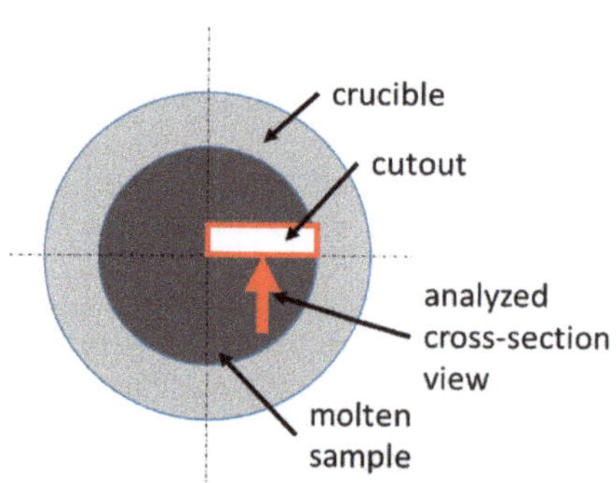

Figure 2. Location of metallographic sample.

The microsections were investigated using a Keyence VHX 7000 digital microscope (Keyence Corporation, Osaka, Japan). The chemical analysis of the gangue layer was performed via SEM-EDX Fei Quanta 200Mk2 (FEI Company, Hillsboro, OR, USA). Small samples were taken from two crucibles, Test 2 with carbonaceous DRI and Test 3 with DRI fines, to analyze carbon content using LECO. Test 3 can be considered representative for the carbon-free DRI tests.

3. Results

Figure 3 shows the voltage as a function of the current for all tests. The different point clouds represent stability regions from different controlling ranges. The more concentrated the points in the diagram are, the more stable the arc is. Since only minor fluctuations could be observed, arc operation was relatively stable in all cases. Nevertheless, the violet Test 3 with the DRI fines shows the broadest range, while the green Test 4 dots are relatively closely distributed. That confirms the arc-stabilizing effect of a slag layer on top of the melt.

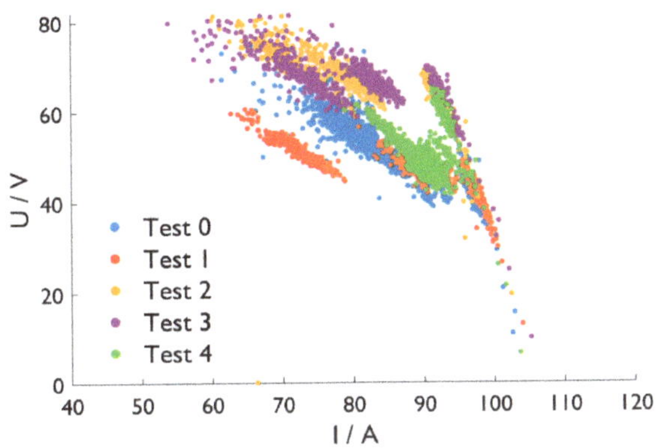

Figure 3. Voltage (U) vs. current (I) diagram.

All relevant furnace parts, the crucible, and the sample materials were weighed for a mass balance. All balances are negative, with Test 2, C-DRI, having the largest mass loss through the gas phase. During this test, 27.0 g dusted and evaporated from the furnace chamber without becoming separated in the off-gas filter. Further, the reduction reactions influence the mass balance in this case. That was the least pronounced at Test 4 with the slag cover and only 2.8 g loss. Tests 1 and 3 are in the middle, with—11.3 and—14.9 g, respectively.

Figure 4 shows in situ photographs during the carbon and slag-free Tests 1 and 3. The electric arc moves over the surface and melts the sponge iron. The pellets in Figure 4a offer a contrast to describe the progress better. A liquid pool is on the left-hand side, while a pellet structure can still be observed on the right side. In contrast, the DRI-fines in Figure 4b do not show any contours, making a detailed observation difficult.

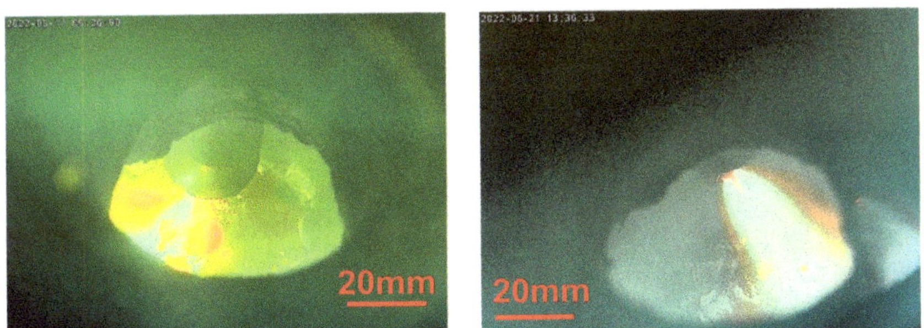

(a) **(b)**

Figure 4. In situ photographs. (**a**) Test 1; (**b**) Test 3.

The formation of bubbles characterizes the carburized sample in Figure 5; see the red circle. These blisters indicate a reduction in residual iron oxide with carbon in the sponge iron sample.

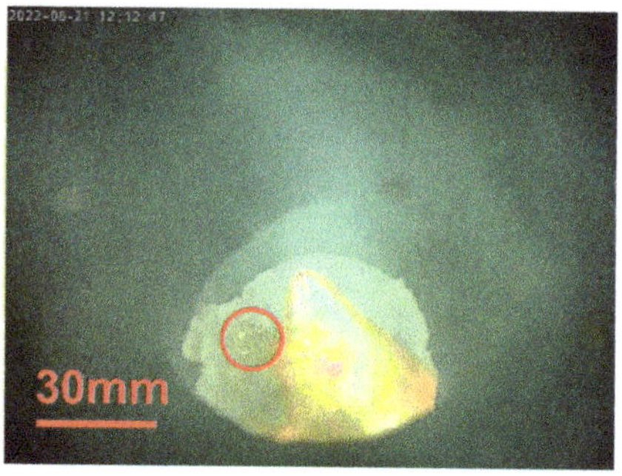

Figure 5. In situ photograph during Test 2.

The dissolution of slag-forming oxide can be observed in Figure 6. The photograph in (a) was taken directly after the addition of powder. In (b), the powder dissolves in the superficial slag layer; see the red circled spot. After charging the oxide powder in (c), the whole surface seems to glow. That differs from the other samples in which the intensive cooling effect through the steel crucible leads to a localized hot spot. This appearance demonstrates the insulating effect of the slag layer.

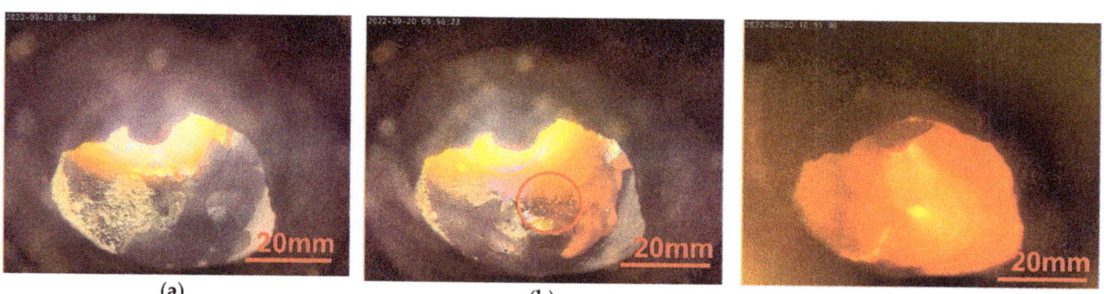

(a) (b) (c)

Figure 6. In situ photographs during Test 4 (**a**) with oxide powder on top; (**b**) dissolution of oxide powder; (**c**) after all slag formers are charged.

Figure 7 shows the off-gas analysis of Test 2 with carburized DRI. The high CO and CO_2 amounts indicate a reduction in residual iron oxide, confirming the observation above. The carbon content of 0.75%, which significantly decreased from the DRI, correlates with the result. A further striking aspect is the increasing hydrogen amount resulting from moisture in the furnace refractory. The higher the temperature, the more moisture is released, which reacts to H_2 and CO after the heterogeneous shift reaction.

Figure 8 provides the off-gas analysis of Test 3 with the DRI fines. H_2 shows an analogous behavior as above; CO and CO_2 result from electrode burn-off as the sample contains no carbon. Unfortunately, the analysis of Test 1 was lost due to a software issue. Nevertheless, Figure 8 is representative of exhaust gas compositions of the carbon-free samples in Tests 1, 3, and 4. Sample 3 remained almost carbon-free with 0.09%, which may have resulted from interactions with the steel crucible or the graphite electrode.

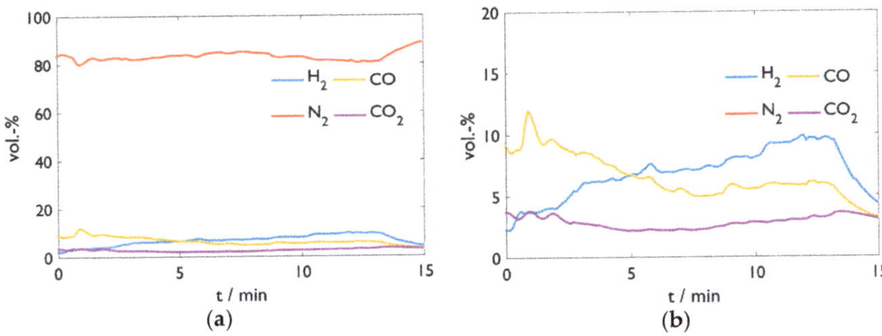

Figure 7. Off-gas analysis of Test 2. (**a**) Overview; (**b**) detailed y-scale.

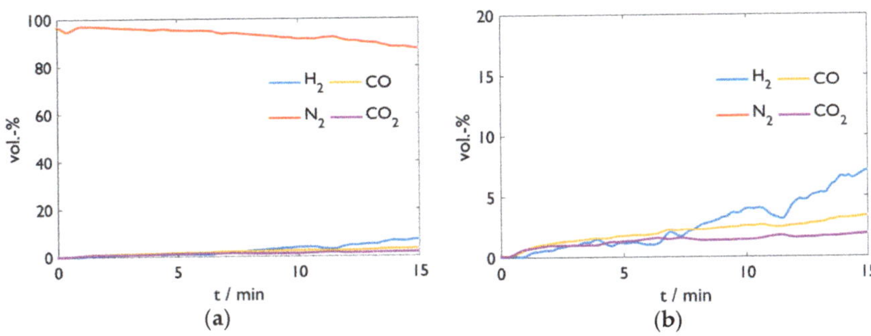

Figure 8. Off-gas analysis of Test 3. (**a**) Overview; (**b**) detailed y-scale.

Figure 9 shows the sample crucibles after the tests. Red dust settles superficially. The most plausible explanation is the formation of iron oxides or hydroxides from the vaporized Fe with moisture from the furnace refractory. That aligns with the slag-covered sample in (d). On the one hand, it shows a primarily grey color; on the other hand, this sample evaporated the least. While the samples in (a) and (b) present a rough surface, the slag layer in (d) is relatively smooth but brittle with many cracks. This layer's dark grey color indicates a high amount of iron oxide. Tests 1 and 3 imply a formation of a concentric structure from the temperature gradient around the electric arc.

Looking at the carburized DRI sample from Test 2 in (b), two aspects catch the eye. Firstly, lots of droplets cover the crucible on the lower right-hand side. Secondly, the sample surface appears less metallic but somewhat yellowish and glassy. A possible explanation for this appearance may be that CO blisters from reduction reactions splash out metallic droplets. These droplets arrive on the crucible surface. Further, the blisters drag the gangue onto the surface of the sample. The splashes also explain the pronounced mass loss, as some could land in the furnace refractory without being noticed.

Figure 10 shows the top layer particle from Test 2 as a digital microscope image from the upper and lower side. While (a) looks metallic with many blisters, (b) seems glassy.

Table 3 provides the approximate composition of the mentioned particle from both sides, determined by SEM-EDX. Although the iron oxide content in the lower part is slightly higher, the compositions do not differ significantly. The shiny gray color indicates that it is partly metalized. Nevertheless, SEM-EDX does not provide the possibility to differentiate between Fe_{met}, Fe^{2+}, and Fe^{3+}.

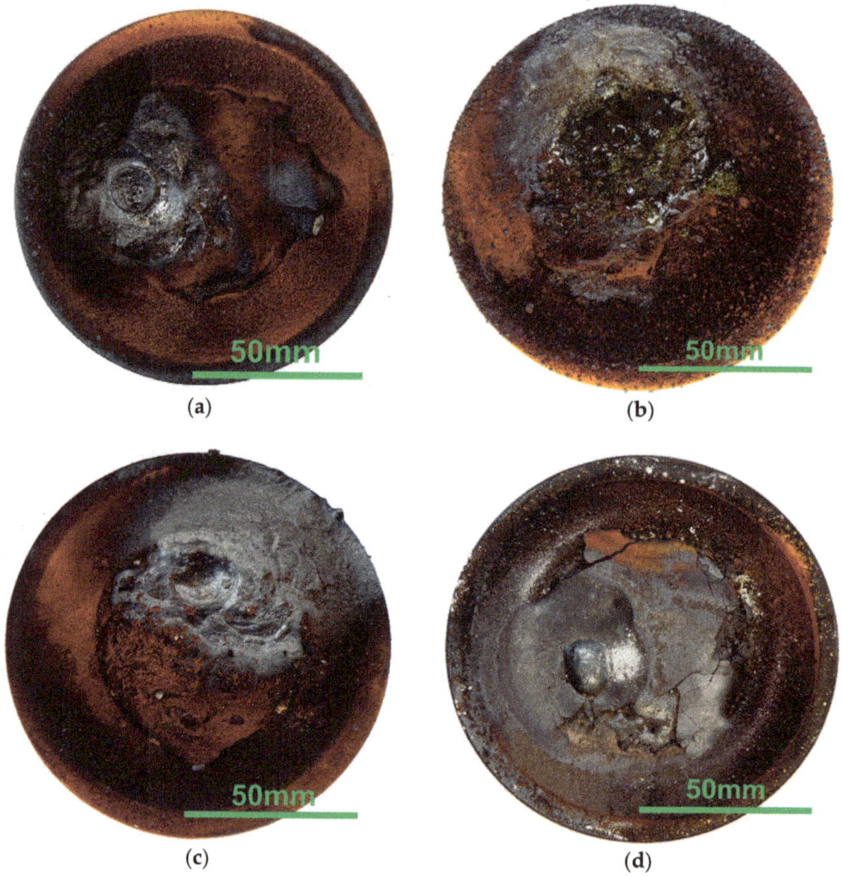

Figure 9. Macroscopic photographs of the sample crucibles after the tests. (**a**) Test 1; (**b**) Test 2; (**c**) Test 3; (**d**) Test 4.

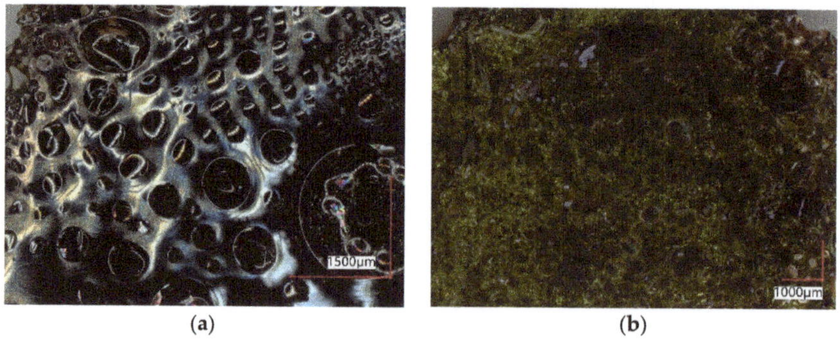

Figure 10. Top layer of Test 2. (**a**) lower side; (**b**) upper side.

Table 3. Chemical composition of Test 2 top layer from Figure 10, determined by SEM-EDX as elements and recalculated as oxides.

wt.-%	FeO	MnO	SiO_2	Al_2O_3	CaO	MgO	TiO_2
Lower side	16.9	9.33	36.3	7.40	19.0	9.57	1.55
Upper side	14.3	9.19	38.3	7.77	19.7	9.16	1.58

Figure 11 shows the digital microscope images of the cross-sections. The following aspects are noticeable:

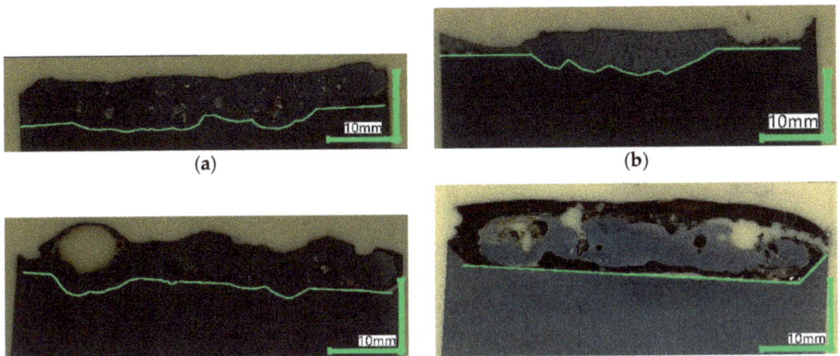

Figure 11. Digital microscope images of cross-sections. (**a**) Test 1; (**b**) Test 2; (**c**) Test 3; (**d**) Test 4.

- The carbon-free samples from Tests 1, 3, and 4 show numerous pores, especially for the fine-DRI sample in (c), with one big blister. Further, they contain dark spots indicative of gangue inclusions.
- The carbon-bearing sample in (b) looks completely different in this case. It forms a dense structure without pores and inclusions. This behavior correlates with the macroscopic observations and the gangue being agglomerated and collected on the sample top.
- Samples in (a), (b) and (c) are merged with the crucible. The opposite is the case in (d). The slag frames the steel; the straight borderline indicates the absence of liquefying of the crucible. This picture is even more striking since the crucible center appears enriched with slag; compare the cutting scheme in Figure 2.

4. Discussion

Small-scale melting tests were performed with different DRI types in the HPSR reactor in the EAF operation. The stable electric arc confirms the good fusibility of all DRI samples. That is especially true if the furnace is operated with a slag layer, leading to little electrical fluctuations, highlighted in the stability areas of Figure 3.

All mass balances are negative, indicating extensive evaporation and dust losses into the furnace chamber without becoming separated in the off-gas filter. That is the most pronounced for the carbon-bearing sample in Test 2. The larger droplets on the steel crucible, visible in Figure 9b, suggest a second effect. CO bubbles from the reduction in residual FeO splash out of the melt into the furnace chamber, carrying liquid droplets. Therefore, this mass loss is not considered in the mass balance. The high CO amount in the off-gas and the low final carbon content of 0.75% indicate significant reduction of residual iron oxide in the DRI.

Test 2 also stands out in another way. The cross-section is free of blisters and gangue inclusions. The gangue accumulates on the surface. This appearance indicates a purifying effect of the CO bubbles in terms of gases and entrapments; the remaining carbon could enhance this effect by lowering the melt viscosity [40].

Test 4 demonstrates the critical influence of slag on the melting process. Besides the thermal insulation effect, the slag also avoids evaporation and dust losses and stabilizes the electric arc slightly. Further, this test demonstrates bath stirring, possibly from electromagnetic forces [41]. While the melt in all other tests merges with the crucible, this is not true in Test 4.

Combining these observations with findings from previous studies, we can explain the crucial importance of the DRI feeding point. For fast DRI melting, a contact between pellet and steel melt is essential [9–12,42]. In the arc center, intensive bath stirring happens. Consequently, besides the high temperature, this mixing effect increases the chance of DRI contacting crude steel, increasing its melting rate. Higher carbon content can further enhance that in the melting stage, which can be decreased during refining before tapping.

5. Conclusions and Outlook

This study demonstrates the importance of the DRI feeding point into the electric arc hotspot. It is not only the high temperature but also the electromagnetic stirring mechanism that leads to an increased chance of DRI contacting the liquid steel and forcing convective heat and material diffusion. Test 4 with slag-forming oxides demonstrates this effect by the formation of a slag layer which covers the steel sample. The slag layer also slightly stabilizes the arc, insulates the bath thermally and prevents severe evaporation.

An increased carbon content during the charging and melting stage seems to be further beneficial. As previous studies show, melting is faster with carbon. Besides that, Test 2 with carbon-bearing DRI suggests that a cleaning effect from the CO bubbles can be expected, which benefits the final crude steel quality. This effect is evident in two respects. On the one hand, the structure of Test 2 appears free of blisters, and on the other hand, the gangue oxides are deposited on the surface instead of being finely dispersed. Last but not least, the fact that more than half of the carbon reacted with residual oxides indicates a high reactivity of carbon in DRI.

In a nutshell, the optimal conditions based on these experiments would be as follows: a highly carburized DRI that is fed directly into the electric arc hot spot. There, the pronounced stirring effect of the electric arc provides the optimal conditions for sponge iron melting. As the sample from carbon-free DRI shows blisters and entrapments, besides the charging spot, the hot heel operation in the EAF also demands critical consideration.

The results are a first step for possible future experiments with this method. Interesting topics could be the comparison of different prereduction states, continuous feed of fine DRI, or quantification of the results by using energy balances and calculating melting rates.

Author Contributions: Conceptualization: A.P. and J.S.; methodology: A.P. and D.E.; validation: J.S. and G.W.; formal analysis: J.S.; resources: H.Z., J.S. and G.W.; writing—original draft preparation: A.P.; writing—review and editing: J.S., D.E., H.Z. and G.W.; visualization: A.P.; supervision: J.S. and G.W.; project administration: G.W.; funding acquisition: J.S. All authors have read and agreed to the published version of the manuscript.

Funding: This research was funded by K1-MET GmbH, metallurgical competence center (funding number FFG No. 869295). The research program of the K1-MET competence center is supported by COMET (Competence Center for Excellent Technologies), the Austrian program for competence centers. COMET is funded by the Federal Ministry for Climate Action, Environment, Energy, Mobility, Innovation, and Technology, the Federal Ministry for Labour and Economy, the provinces of Upper Austria, Tyrol, and Styria, and the Styrian Business Promotion Agency (SFG). In addition, this research work was partially financed by the industrial partners Primetals Technologies Austria GmbH, voestalpine Stahl GmbH, voestalpine Stahl Donawitz GmbH and thyssenkrupp Steel Europe AG, and the scientific partner Montanuniversität Leoben.

Data Availability Statement: The data presented in this study are available within the article.

Conflicts of Interest: The authors declare no conflict of interest.

References

1. World Steel in Figures. Available online: https://worldsteel.org/steel-topics/statistics/world-steel-in-figures/ (accessed on 23 September 2022).
2. Lule, R.; Lopez, F.; Espinoza, J.; Torres, R.; Morales, G. The Experience of ArcelorMittal Lázaro Cardenas Flat Carbon. *Direct Midrex* **2009**, *3*, 3–8.
3. Bayer, F.; Strasser, J.; Trenkler, H. Hadeed's new flat-steel mill at Al-Jubail-from Mini to Midi. *Steel Times* **2000**, *228*, 364.
4. Jung, H.; Al-lbrahim, N.; Al-Sayegh, A.; Kaspar, S.; Pirklbauer, W. Hadeed-concept and latest results of the world's largest EAF plant for long products on DRI-basis. *Rev. Métallurgie* **1994**, *91*, 1123–1130. [CrossRef]
5. DRI Technology at Emirates Steel | Tenova. Available online: https://tenova.com/newsroom/latest-tenova/dri-technology-emirates-steel (accessed on 24 January 2023).
6. Midrex Technologies, Inc. World DRI Production Reaches 119.2 Mt in 2021 Midrex Publishes World Direct Reduction Statistics. Available online: https://www.midrex.com/company-news/world-dri-production-reaches-119-2-mt-in-2021-midrex-publishes-world-direct-reduction-statistics/ (accessed on 28 September 2022).
7. Eder, W. Environment–Climate–Energy: Quo Vadis, Industry? *Berg Huettenmaenn. Mon.* **2017**, *162*, 494–497. [CrossRef]
8. Griesser, A.; Buergler, T. Use of HBI in Blast Furnace. *Berg Huettenmaenn. Mon.* **2019**, *164*, 267–273. [CrossRef]
9. González, O.J.P.; Ramírez-Argáez, M.A.; Conejo, A.N. Mathematical Modeling of the Melting Rate of Metallic Particles in the Electric Arc Furnace. *ISIJ Int.* **2010**, *50*, 9–16. [CrossRef]
10. Pineda-Martínez, E.; Hernández-Bocanegra, C.A.; Conejo, A.N.; Ramirez-Argaez, M.A. Mathematical Modeling of the Melting of Sponge Iron in a Bath of Non-Reactive Molten Slag. *ISIJ Int.* **2015**, *55*, 1906–1915. [CrossRef]
11. Ramirez-Argaez, M.A.; Conejo, A.N.; López-Cornejo, M.S. Mathematical Modeling of the Melting Rate of Metallic Particles in the EAF under Multiphase Flow. *ISIJ Int.* **2015**, *55*, 117–125. [CrossRef]
12. Pfeiffer, A.; Wimmer, G.; Schenk, J. Investigations on the Interaction Behavior between Direct Reduced Iron and Various Melts. *Materials* **2022**, *15*, 5691. [CrossRef]
13. Penz, F.; Schenk, J.; Ammer, R.; Klösch, G.; Pastucha, K. Dissolution of Scrap in Hot Metal under Linz–Donawitz (LD) Steelmaking Conditions. *Metals* **2018**, *8*, 1078. [CrossRef]
14. Penz, F.M.; Schenk, J. A Review of Steel Scrap Melting in Molten Iron-Carbon Melts. *Steel Res. Int.* **2019**, *90*, 1900124. [CrossRef]
15. Penz, F.M.; Schenk, J.; Ammer, R.; Klösch, G.; Pastucha, K.; Reischl, M. Diffusive Steel Scrap Melting in Carbon-Saturated Hot Metal-Phenomenological Investigation at the Solid-Liquid Interface. *Materials* **2019**, *12*, 1385. [CrossRef] [PubMed]
16. Hornby, S.; Madias, J.; Torre, F. Myths and realities of charging DRI/HBI in electric arc furnaces. *Iron Steel Technol.* **2016**, 81–90. [CrossRef]
17. Kirschen, M.; Hay, T.; Echterhof, T. Process Improvements for Direct Reduced Iron Melting in the Electric Arc Furnace with Emphasis on Slag Operation. *Processes* **2021**, *9*, 402. [CrossRef]
18. Hornby, S.; Madias, J.; Torre, F. DRI/HBI-exploding the myths. *Steel Times Int.* **2016**, *40*, 24.
19. Omar, A.M.; Appasamy, T.A.; Memoli, F. DC EAF with high DRI feeding rates through multipoint injection. *Metall. Plant Technol. Int.* **2004**, *27*, 58–67.
20. Al Dhaeri, A.; Razza, P.; Patrizio, D. Excelent operating results of the integrated minimill #1 at Emirates Steel Industries: Danieli. *MPT Int.* **2010**, *33*, 34–40.
21. Heo, J.; Park, J.H. Interfacial reactions between magnesia refractory and electric arc furnace (EAF) slag with use of direct reduced iron (DRI) as raw material. *Ceram. Int.* **2022**, *48*, 4526–4538. [CrossRef]
22. Oh, M.K.; Kim, T.S.; Park, J.H. Effect of CaF2 on Phosphorus Refining from Molten Steel by Electric Arc Furnace Slag using Direct Reduced Iron (DRI) as a Raw Material. *Met. Mater. Trans B* **2020**, *51*, 3028–3038. [CrossRef]
23. Heo, J.H.; Park, J.H. Effect of Direct Reduced Iron (DRI) on Dephosphorization of Molten Steel by Electric Arc Furnace Slag. *Met. Mater. Trans. B* **2018**, *49*, 3381–3389. [CrossRef]
24. Heo, J.H.; Park, J.H. Assessment of Physicochemical Properties of Electrical Arc Furnace Slag and Their Effects on Foamability. *Met. Mater. Trans. B* **2019**, *50*, 2959–2968. [CrossRef]
25. Hay, T.; Visuri, V.-V.; Aula, M.; Echterhof, T. A Review of Mathematical Process Models for the Electric Arc Furnace Process. *Steel Res. Int.* **2021**, *92*, 2000395. [CrossRef]
26. Hay, T.; Echterhof, T.; Visuri, V.-V. Development of an Electric Arc Furnace Simulator Based on a Comprehensive Dynamic Process Model. *Processes* **2019**, *7*, 852. [CrossRef]
27. Kieush, L.; Schenk, J.; Koveria, A.; Rantitsch, G.; Hrubiak, A.; Hopfinger, H. Utilization of Renewable Carbon in Electric Arc Furnace-Based Steel Production: Comparative Evaluation of Properties of Conventional and Non-Conventional Carbon-Bearing Sources. *Metals* **2023**, *13*, 722. [CrossRef]
28. Echterhof, T. Review on the Use of Alternative Carbon Sources in EAF Steelmaking. *Metals* **2021**, *11*, 222. [CrossRef]
29. Kieush, L.; Rieger, J.; Schenk, J.; Brondi, C.; Rovelli, D.; Echterhof, T.; Cirilli, F.; Thaler, C.; Jaeger, N.; Snaet, D.; et al. A Comprehensive Review of Secondary Carbon Bio-Carriers for Application in Metallurgical Processes: Utilization of Torrefied Biomass in Steel Production. *Metals* **2022**, *12*, 2005. [CrossRef]
30. Hanel, M. Characterization of Ferrous Burden Material for Use in Ironmaking Technologies. Ph.D. Thesis, Montanuniversität Leoben, Leoben, Austria, 2014.

31. Bhattacharyya, A.; Schenk, J.; Jäger, M.; Stocker, H.; Thaler, C. Experimental Simulation of the Interaction of Slag and Hot Metal with Coke at the Bosh Region of Blast Furnace. *Berg Huettenmaenn. Mon.* **2017**, *162*, 28–33. [CrossRef]
32. ISO 11258:2015. ISO. Available online: https://www.iso.org/standard/62145.html (accessed on 13 May 2023).
33. Spreitzer, D. Development of Characterization Methods for the Evaluation of the Kinetic Behavior and the Fluidization of Iron Ore Fines during Hydrogen-Induced Fluidized Bed Reduction. Ph.D. Thesis, University of Leoben, Leoben, Austria, 2021.
34. Spreitzer, D.; Schenk, J. Iron Ore Reduction by Hydrogen Using a Laboratory Scale Fluidized Bed Reactor: Kinetic Investigation—Experimental Setup and Method for Determination. *Met. Mater. Trans. B* **2019**, *50*, 2471–2484. [CrossRef]
35. Ernst, D.; Zarl, M.A.; Cejka, J.; Schenk, J. A New Methodological Approach on the Characterization of Optimal Charging Rates at the Hydrogen Plasma Smelting Reduction Process Part 2: Results. *Materials* **2022**, *15*, 4065. [CrossRef]
36. Zarl, M.A.; Ernst, D.; Cejka, J.; Schenk, J. A New Methodological Approach to the Characterization of Optimal Charging Rates at the Hydrogen Plasma Smelting Reduction Process Part 1: Method. *Materials* **2022**, *15*, 4767. [CrossRef]
37. Zarl, M.A.; Farkas, M.A.; Schenk, J. A Study on the Stability Fields of Arc Plasma in the HPSR Process. *Metals* **2020**, *10*, 1394. [CrossRef]
38. Ernst, D.; Zarl, M.A.; Farkas, M.A.; Schenk, J. Effects of the Electrodes' Shape and Graphite Quality on the Arc Stability During Hydrogen Plasma Smelting Reduction of Iron Ores. *Steel Res. Int.* **2023**, 2200818. [CrossRef]
39. Ernst, D.; Manzoor, U.; Souza Filho, I.R.; Zarl, M.A.; Schenk, J. Impact of Iron Ore Pre-Reduction Degree on the Hydrogen Plasma Smelting Reduction Process. *Metals* **2023**, *13*, 558. [CrossRef]
40. Feng, G.; Jiao, K.; Zhang, J.; Gao, S. High-temperature viscosity of iron-carbon melts based on liquid structure: The effect of carbon content and temperature. *J. Mol. Liq.* **2021**, *330*, 115603. [CrossRef]
41. Ramírez, M.; Alexis, J.; Trapaga, G.; Jönsson, P.; Mckelliget, J. Mathematical Modeling of Iron and Steel Making Processes. Modeling of a DC Electric Arc Furnace. Mixing in the Bath. *ISIJ Int.* **2001**, *41*, 1146–1155. [CrossRef]
42. Gonzalez, O.J.P.; Ramírez-Argáez, M.A.; Conejo, A.N. Effect of Arc Length on Fluid Flow and Mixing Phenomena in AC Electric Arc Furnaces. *ISIJ Int.* **2010**, *50*, 1–8. [CrossRef]

metals

MDPI

Review

Reoxidation Behavior of the Direct Reduced Iron and Hot Briquetted Iron during Handling and Their Integration into Electric Arc Furnace Steelmaking: A Review

Lina Kieush [1,*], Stefanie Lesiak [1], Johannes Rieger [1], Melanie Leitner [1], Lukas Schmidt [1] and Oday Daghagheleh [2]

1 K1-MET GmbH, Stahlstrasse 14, 4020 Linz, Austria
2 Chair of Ferrous Metallurgy, Montanuniversität Leoben, Franz-Josef-Str. 18, 8700 Leoben, Austria
* Correspondence: lina.kieush@k1-met.com

Abstract: This paper studies the integration of direct reduced iron (DRI) and hot briquetted iron (HBI) into the steelmaking process via an electric arc furnace (EAF). Considering a variety of DRI production techniques distinguished by different reactor types, this paper provides a comparative overview of the current state. It delves into significant challenges, such as the susceptibility of DRI to reoxidation and the necessity of thorough handling to maintain its quality. The effectiveness of several reoxidation mitigation strategies, including the application of thin oxide layers, briquetting, various coatings, and nitride formation in ammonia-based reduction processes, is evaluated. Most existing studies have primarily focused on the reoxidation of DRI rather than on HBI, despite the fact that HBI may undergo reoxidation. The importance of DRI/HBI in offering an alternative to the integrated steelmaking route is highlighted, focusing on how it changes the EAF process compared to those for melting scrap. This paper also identifies several research prospects for further DRI/HBI applications in steel production.

Keywords: direct reduced iron; hot briquetted iron; reoxidation; electric arc furnace; steelmaking

1. Introduction

The direct reduction (DR) of iron ore (in the form of lumps, pellets, or fines) is a process of forming low-valence oxides in the production of direct reduced iron (DRI), also known as sponge iron [1,2], where oxygen is removed from iron ore below the melting point of reactants and products, applying different reductants, namely from the conversion of natural gas (NG), syngas [3], byproduct gases [4,5], coal [6], hydrogen (H_2) [7], etc. The emissions intensity of DR varies with the chosen reductant, driven by factors such as gas accessibility, costs, and the requirement for high-grade iron ore. High-grade ore is important for DRI technology, yet it makes up less than 5% of the global iron ore supply [8]. Coal-based DRI reduces carbon dioxide (CO_2) emissions by 38%, contrasted with blast furnace-basic oxygen furnace (BF-BOF) ironmaking. Using a combination gas of methane (CH_4), H_2, and carbon monoxide (CO) for DRI reduces CO_2 emissions by 61% as opposed to BF-BOF [9]. In comparison with the NG-based DR process, using H_2 can reduce up to 91% of directly emitted CO_2, not accounting for the CO_2 intensity of the additional electricity required [10]. However, this value also hinges on the method used for H_2 production and its associated CO_2 footprint.

Different reactor technologies are used for DRI production, such as shaft furnaces, rotary kiln (RK), rotary hearth furnaces (RHF), and fluidized bed reactors. However, 96.8% of the DRI is produced using two technologies, while 72.4% is attributed to shaft furnaces and 24.4% to rotary kilns (RK) [11]. In 2022, the worldwide production of DRI reached 127.36 million tons (Mt), marking an increase of nearly 6.9% over the prior record of 119.2 Mt established in 2021. Within the spectrum of DRI approaches, MIDREX® led with approximately 57.8% of the global DRI output for the year. In comparison, other gas-based technologies contributed 12.1% from HYL/ENERGIRON® and 2.2% from PERED. The remaining production primarily involved coal-based DR processes in RK, accounting for 27.9% [12].

DRI is an alternative for the iron production industry [13,14]. In addition, DRI can be compacted to reduce porosity and converted to hot briquetted iron (HBI) [15]. This process not only prevents reoxidation but also enables HBI to be transported over long distances and stored outside for long periods. However, cases of reoxidation of HBI might also occur.

The electric arc furnace (EAF) is well-suited to integrating DRI/HBI owing to its flexibility in charge materials. EAF can melt steel scrap, DRI/HBI, or a combination of both. With the utilization of scrap in the BF-BOF and EAF steelmaking processes, the available supply of scrap metal is becoming more limited [15,16].

In the EAF, since DRI has a higher bulk density than most steel scrap types, surpassing slag density, this characteristic promotes its melting at the interface between slag and metal. The remaining wuestite (FeO) content in DRI interacts with the carbon in the liquid metal, enhancing the formation of foaming slag that protects the refractory lining from the electric arcs. Nonetheless, controlling the rate at which DRI is introduced into the EAF is important. The feeding rate varies, influenced by the chemical composition and metallization degree of DRI, the temperature of the metal bath, an initial amount of liquid steel (LS) bath, and the mixing energy from the oxygen-carbon injectors along with the stirring [17]. This rate typically falls within a range of 27–35 $kg{\cdot}(min{\cdot}MW)^{-1}$ [16]. Most DRI/HBI can be melted with a varying share of steel scrap in EAF. The specific CO_2 emissions are within the range of 0.9–1.8 $t_{CO2}{\cdot}t_{LS}{}^{-1}$, including local CO_2 intensity of electrical energy for EAF [18,19].

Furthermore, there are other perspectives, including flexibility, quality control, and resource optimization, that should be considered, including the important environmental aspect. Adding DRI to the EAF allows for more precise control over the final composition of steel. This is necessary for producing high-quality steel grades that meet strict specifications for alloy content and impurity levels. Additionally, using DR might enable the efficient use of iron ore resources, including low-grade iron ores unsuitable for some metallurgical processes. Although applying the proportion of DRI or HBI for EAF is attractive, it may pose issues that should be addressed.

The focus of this paper lies in considering and showing the primary outcomes and aspects associated with DRI or HBI, starting with the characteristics and comparison of available main types of reduction reactors. This study evaluates the current research state, focusing on challenges such as reoxidation of DRI or HBI that may arise post-production and exploring potential solutions for handling and transportation. To the best of our knowledge, existing review papers related to the topic of DRI and HBI have not focused on the challenges related to reoxidation, revealing the necessity for further research. Moreover, the paper discusses the research results on applying DRI or HBI into EAF under laboratory and industrial scale from the process-wise perspective, assesses their impact and limitations, and suggests areas for future research.

The review is organized as follows: Section 2 presents the main principles, routes, and reactors comparison of the DR process, along with the resulting DRI characteristics; Section 3 discusses the carburization phenomenon that might occur during or after the reduction process; Section 4 focuses on the reoxidation issue and summarises the current state of the existing methods for mitigating this effect; Section 5 contains generalized measurements concerning storage, handling, and transportation; Section 6 comprises current research results concerning DRI or HBI application in an EAF steelmaking, considering the effect on this process; and Section 7 presents summary and research prospects.

2. Direct Reduction of Iron Ore

2.1. Main Principles of DR

The principles of the DR of iron ore have been outlined [20–22], and several points can be mentioned. The initial stage of reduction typically involves hematite (Fe_2O_3), which is first converted into magnetite (Fe_3O_4), followed by a reduction to FeO. Iron oxide reduction shows a broad range of possible reactions before its final reduction to metallic iron, according to Equations (1)–(8). FeO frequently shows non-stoichiometry, characterized by a cation deficiency represented as $Fe_{(1-x)}O$. Nevertheless, to simplify the reaction equations (specifically Equations (2), (3), (6) and (7)), FeO is depicted without considering the precise stoichiometric ratio. The process of DR in a CO atmosphere often results in carbon deposition, a consequence of the inverse Boudouard reaction (CO disproportionation reaction) at temperatures below 1000 °C, along with the Beggs reaction [23,24], according to Equations (9) and (10). The Beggs reaction and the CO disproportionation reaction, both exothermic, generate H_2O or CO_2. This leads to a decrease in the quality of the gas within the reactor, thereby making the conditions less favorable for reduction processes [25].

$$3Fe_2O_3\ (s) + CO\ (g)\ \leftrightarrow\ 2Fe_3O_4\ (s) + CO_2\ (g),\ \Delta H^{\theta}_{298K} = -43\ kJ{\cdot}mol^{-1}, \quad (1)$$

$$Fe_3O_4\ (s) + CO\ (g)\ \leftrightarrow\ 3FeO\ (s) + CO_2\ (g),\ \Delta H^{\theta}_{298K} = 19\ kJ{\cdot}mol^{-1}, \quad (2)$$

$$FeO\ (s) + CO\ (g)\ \leftrightarrow\ Fe\ (s) + CO_2\ (g),\ \Delta H^{\theta}_{298K} = -11\ kJ{\cdot}mol^{-1}, \quad (3)$$

$$Fe_2O_3\ (s) + 3CO\ (g)\ \leftrightarrow\ 2Fe\ (s) + 3CO_2\ (g),\ \Delta H^{\theta}_{298K} = -23\ kJ{\cdot}mol^{-1}, \quad (4)$$

$$3Fe_2O_3\ (s) + H_2\ (g)\ \leftrightarrow\ 2Fe_3O_4\ (s) + H_2O\ (g),\ \Delta H^{\theta}_{298K} = -2\ kJ{\cdot}mol^{-1}, \quad (5)$$

$$Fe_3O_4\ (s) + H_2\ (g)\ \leftrightarrow\ 3FeO\ (s) + H_2O\ (g),\ \Delta H^{\theta}_{298K} = 60\ kJ{\cdot}mol^{-1}, \quad (6)$$

$$FeO\ (s) + H_2\ (g)\ \leftrightarrow\ Fe\ (s) + H_2O\ (g),\ \Delta H^{\theta}_{298K} = 30\ kJ{\cdot}mol^{-1}, \quad (7)$$

$$Fe_2O_3\ (s) + 3H_2\ (g)\ \leftrightarrow\ 2Fe\ (s) + 3H_2O\ (g),\ \Delta H^{\theta}_{298K} = 100\ kJ{\cdot}mol^{-1}, \quad (8)$$

$$2CO\ (g)\ \leftrightarrow\ C\ (s) + CO_2\ (g),\ \Delta H^{\theta}_{298K} = -172\ kJ{\cdot}mol^{-1}, \quad (9)$$

$$CO\ (g) + H_2\ (g)\ \leftrightarrow\ C\ (s) + H_2O\ (g),\ \Delta H^{\theta}_{298K} = -131\ kJ{\cdot}mol^{-1}, \quad (10)$$

The metallization degree and the behavior of iron ore change based on the reducing atmosphere, temperature, and reduction rate [24]. The reduction process of Fe_2O_3 becomes unstable at temperatures under 570 °C [26]. At temperatures above 570 °C, the initial reduction phase (from Fe_2O_3 to Fe_3O_4) necessitates a mildly reducing atmosphere and is typically seen as an irreversible process [27]. Subsequent transformation from Fe_3O_4 to FeO is facilitated by both CO and H_2, and this step is endothermic. Below 570 °C, FeO does not form, leading to the DR of Fe_3O_4 to metallic iron.

Generally, the overall reduction of Fe_2O_3 by CO is exothermic, in contrast to the endothermic nature of H_2 reduction [28,29]. Typically, the highly endothermic nature of reduction systems necessitates the addition of extra energy to maintain a steady reduction temperature when using H_2.

The transition from FeO to Fe, the most challenging reaction, demands a highly reducing environment. Beyond 810 °C, the reducing capability of hydrogen surpasses that of CO [30], influenced by kinetic factors and the water-gas shift reaction. In addition, the reaction rate would increase with the higher H_2 content at temperatures above 1000 °C [31]. The reduction rates with H_2 are much faster, up to 10 times than those with CO [23,32,33]. Additionally, H_2 has a higher diffusion rate and could penetrate much deeper into the crystal structure of the iron oxide, leading to a higher reduction rate and a greater metallization degree [24].

From the kinetic perspective, reduction in shaft furnaces involves several sequential steps, with the overall rate being determined by the slowest of these steps. In a shaft furnace, reduction contains the flow of gas and solid phases, the exchange of mass and heat between these phases, and the reactions occurring at the interface between gas and solid. In general, these steps encompass the movement of gases to the reaction site, the chemical reaction itself, and removing reaction products from the site. Additionally, solid-state diffusion can play a role, such as when FeO is being reduced to iron and a gradient of ferrous iron is created, necessitating the movement of ferrous ions to the reaction site for further conversion. Notably, temperature, pressure, flow rate, reducing gas composition, as well as size, morphology, free surface area, porosity, tortuosity, and chemical composition [23,34] of the materials involved are factors influencing the kinetics of the mentioned reactions. The reduction temperature varies, falling within the 850–960 °C range [26,35], and operating pressures utilizing NG range from 0.23 to 0.60 MPa [26].

Iron oxides are reduced across multiple stages in fluidized bed reactors, representing the counter-current flow of gas and solids seen in shaft furnaces. Initially, both the solids and reductant gas are heated in a preheater, ensuring they reach the first reduction chamber at temperatures conducive to offsetting the endothermic nature of the reduction process. The reduction usually occurs at temperatures ranging from 760 °C [36], when the DRI product may become pyrophoric, to 800 °C, where issues like sticking and de-fluidization might arise [37]. Operating pressures utilizing NG range from 1.1 to 1.3 MPa [26].

In RHF and RK, despite the occurrence of similar reactions, the mechanisms and rate-limiting steps differ. Reduction predominantly occurs within individual pellets that possess sufficient carbon to facilitate the reduction of iron oxide. RHF functions at higher temperatures of around 1300 °C [38] and typically features shorter residence times than RK, which operates at about 960–1100 °C [39] and atmospheric pressure. According to the study [38], achieving the highest reduction efficiency at 1250 °C for RHF requires an optimal carbon/Fe_2O_3 molar ratio of 1.66. Additionally, pellets reduced under these conditions demonstrated enhanced compressive strength.

2.2. Main Routes and Reactors for the DR Processes

The selection of raw materials optimal for DR in steelmaking is based on several principles: their chemical and physical properties, their ability to undergo reduction, and the overall cost-effectiveness encompassing both the DR process and subsequent steelmaking [40]. The DR process utilizes NG, which typically operates with vertical shaft furnaces and fluidized bed reactors (Figure 1). Meanwhile, coal, either directly or after the conversion process into syngas, can be utilized in DR processes in various systems, including vertical shaft furnaces [41], fluidized bed reactors [42,43], RHF [44], and RK [45,46].

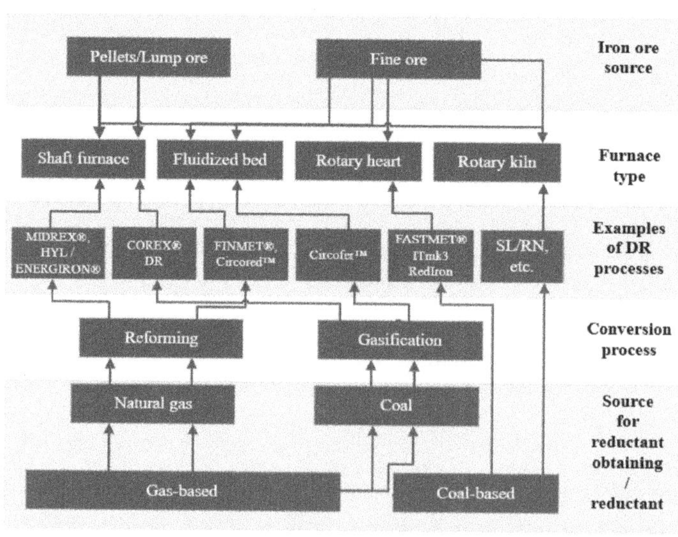

Figure 1. General overview of the main routes of direct reduction based on reductant and furnace type, adapted from Ref. [47].

The majority of DRI is produced in a shaft furnace [48,49], which necessitates a uniformly coarse feed. A shaft furnace is employed for processing pellets or lump ore.

Given the high velocity of gases and the abrasive environment within shaft furnaces, fine particles are considered unsuitable for feedstock; such particles are likely to be removed by the gas stream, requiring subsequent collection and recirculation. Fluidized bed DR processes are exceptions, where iron ore, refined via ore-beneficiation, results in iron ore concentrate powder, which can be directly used [50]. Assuming a substantial number of studies have focused on examining and analyzing reactor types for DR, Table 1 provides a comparative overview of the primary DR reactors, as well as their main advantages and disadvantages.

Table 1. Comparison of main DR reactors, according to [9,15,51–57].

Furnace Type	Form of Reactant (Iron Ore)	Reductant	Short Description of the Process	Advantages	Disadvantages	Metallization Degree of the Product
Shaft furnace	Lump, pellets	Converted NG, syngas, H_2	A vertical furnace in which iron ore pellets or lumps are charged at the top, and reducing gas flows upward, reducing the iron oxides to iron. It relies on solid-gas reactions at high temperatures.	Large production capacity. High product quality.	Ore in pellet form should have a minimum of 67% Fe content. High cost of reducing gas.	≥92%
Fluidized bed	Fines	Converted NG, syngas, H_2	Iron ore particles are suspended in an upward-flowing stream of reducing gas, promoting high mass and heat transfer. Using a fluidizing medium enhances the reaction kinetics by increasing the contact surface area between the reducing gas and the iron ore particles.	Iron ore fines can be used without pelletizing and agglomeration. Fast reaction rate.	Complex structure of reactors. Particle sticking tendency.	≥92%
Rotary kiln	Lump, pellets	Coal (5–20 mm) and recycled char	A kiln is a rotating cylindrical vessel through which lumps or pellets, along with coal or gas as a reductant, are passing. The kiln is slightly inclined to assist material flow. Reduction occurs in solid and gaseous phases; the rotating action promotes mixing and contact between ore and reducing agents, enhancing the reduction process.	Various types of coal can be used. Processing low-grade ores. Flexible structure of the reactor.	Significant levels of impurities, including sulfur. Slow reaction rate. Significant energy usage. Problems with kiln lining agglomeration.	≥90%
		Char, coal				
Rotary hearth furnace	Composite pellets (16–22 mm)	Coal (70% below 45 µm)	A flat, refractory-lined furnace that rotates, carrying the iron ore and carbon mixture through different temperature zones for reduction. Direct reduction occurs in a solid state, with heat supplied predominantly by radiation from overhead burners, facilitating rapid heating and reduction.	Processing low-grade ores or iron-containing by-products. Simple structure of furnace.	Significant levels of impurities, including sulfur. High CO_2 emissions.	≥93%
	Composite pellets	Coal				

2.3. DRI Characteristics

Firstly, the characteristics of the feedstock for DRI pellets should be briefly considered. The main characteristics are presented in Table 2, and the iron content should exceed 67% [58,59]. A decreased iron content, leading to a higher proportion of gangue, can create operational challenges in the EAF. These challenges encompass more electricity usage and increased wear on the refractory lining of the furnace, attributed to a rise in slag volume and the transfer of iron to the slag in the form of FeO [60]. Nevertheless, the impact of DRI application in the EAF steelmaking process will be discussed further in Section 6.

Table 2. Characteristics of the main DR pellets, adapted from Refs. [51,58].

Characteristics	Value
Fe_{tot}, wt.%	≥67.0
SiO_2, wt.%	1.0–3.0
Al_2O_3, wt.%	0.2–3.0
MgO, wt.%	0.2–0.9
CaO, wt.%	0.4–1.2
Sulfur, wt.%	≤0.008
Phosphorus, wt.%	≤0.03
Pellet size, mm	9.0–16.0
Porosity, %	~50.0
Compression strength, N	2500–3000
Tumble index, % + 6.15 mm	92.0–95.0
Reducibility index, %	92.0–95.0
Density, $t{\cdot}m^{-3}$	
bulk	1.6–1.9
apparent	~3.5

Alkali levels should be minimized to prevent swelling and degradation throughout the reduction process. Phosphorus content should be kept to the lowest, below 0.03 wt.%. For DR pellets intended for use in the shaft furnace, sulfur content should be maintained below 0.008 wt.% to avoid clogging the reformer tubes [61].

The quality of DRI (Table 3) is determined by two categories of characteristics: main properties resulting from the DR process, like the metallization degree and carbon content, and inherent properties unrelated to the process, such as gangue content in the raw material. In general, DRI is produced mainly in quality according to the definition by the International Iron Metallics Association [62], characterized by a total iron content ranging between 86.1 and 94.0 wt.% [63] and a metallization degree of 92.0 to 96.0 wt.% [64]. Sulfur content in DRI should be within the range of 0.001–0.03 wt.%, and the phosphorus content should be limited to 0.001–0.09 wt.%. Trace elements like copper, nickel, chromium, molybdenum, and tin are negligible [40].

Table 3. Selected characteristics of DRI, adapted from Refs. [19,62,65–69].

Characteristics	Range, wt.%
Metallization degree	92.0–96.0
Fe_{tot}	86.1–94.0
Fe_{metal}	81.0–89.0
Carbon content	0.02 *–4.5
Sulfur content	0.001–0.03
Phosphorus content	0.001–0.09
Gangue	
Acid	2.5–7.6
Basic	0.5–2.7

* H_2 of the chemical grade was used as a reductant.

DRI predominantly contains silica as the primary gangue component, maintaining an acid gangue composition (mainly SiO_2, Al_2O_3, TiO_2, P_2O_5, Cr_2O_3) and containing basic gangue, for instance, CaO, MgO, MnO, etc. [70,71]. Typically, high levels of gangue elements and inert oxides can diminish the strength of pellets [72]. Oxides, such as SiO_2, Al_2O_3, TiO_2, and V_2O_5, require neutralization through lime additions, influencing both the volume and characteristics of the slag in EAF [73]. The impact of gangue oxides SiO_2, CaO, MnO, and MgO on the reduction process is multifaceted. Although impurities like CaO, MnO, and MgO can lead to swelling and cracking, the right amounts of these gangues could be advantageous. They can create new pathways for the movement of reactants and products, thereby enhancing the rate of reduction reactions [72]. Additionally, according to Tokuda et al. [74], adding SiO_2 to iron oxides significantly reduces their reducibility due to the formation of silicates during the reduction process. This silicate formation leads to low porosity, further decreasing the reducibility of the iron ores. MgO interacts with the intermediate FeO to form (Mg, Fe)O, a compound that presents challenges for reduction [75]. These play a role in prohibiting the reduction process. From the perspective of an EAF, SiO_2 is a surface-active oxide that initially presents a high content in the slag of the hot heel because fluxes are not added until the heat begins. According to Gonzalez et al. [76], this high SiO_2 content can slow the FeO reduction rate. However, this slowing effect can be beneficial as it prolongs the reduction reaction while maintaining an effective, though lower, foam height. As the slag becomes more acidic, the adsorption of SiO_2 at the gas-slag interface intensifies. This reduces the interfacial area available for FeO reduction from the slag.

Table 4 outlines the physical properties of DRI. The density of DRI fluctuates based on its specific form and composition, ranging between 1.5 and 1.9 $t \cdot m^{-3}$ for bulk density and between 3.2 and 3.6 $t \cdot m^{-3}$ for apparent density. The volumetric porosity, however, varies with the production technique and has a mean value of approximately 47.0%. The interaction of DRI with moisture is an important consideration; DRI tends to absorb moisture (12.0–15.0%) from its surroundings, which can lead to oxidation and consequent quality degradation.

Table 4. DRI physical properties, adapted from Refs. [35,58,62,64,77,78].

Characteristics	Value
Bulk density, $t \cdot m^{-3}$	1.5–1.9
Apparent density, $t \cdot m^{-3}$	3.2–3.6
Specific surface area, $m^2 \cdot g^{-1}$	0.5–4.0
Volumetric porosity, vol.%	~47.0
Average size, mm	4.0–20.0
Weight, g	3.0–4.0
Water absorption (saturated), %	12.0–15.0
Fines (−4 mm), %	~5.0

The issue commonly linked to the high specific surface area, porosity, and low density of DRI can often be mitigated through the briquetting process used to produce HBI. This will be detailed in Section 4.3.

3. Carburization

Considering the several advantages of the EAF that are partially associated with the carbon content of DRI [10,79], discussing the carburization phenomenon is relevant. Carburization of DRI during and after the reduction process has various purposes and can be reflected via several reactions, according to Equations (11)–(15):

$$3Fe + CH_4 \rightarrow Fe_3C + 2H_2,\ \Delta H^{\theta}_{298K} = 97\ kJ \cdot mol^{-1}, \tag{11}$$

$$3Fe + 2CO \rightarrow Fe_3C + CO_2,\ \Delta H^{\theta}_{298K} = -150\ kJ \cdot mol^{-1}, \tag{12}$$

$$3Fe + CO + H_2 \leftrightarrow Fe_3C + H_2O,\ \Delta H^{\theta}_{298K} = -108\ kJ \cdot mol^{-1}, \tag{13}$$

$$3FeO + 5CO \leftrightarrow Fe_3C + 4CO_2,\ \Delta H^{\theta}_{298K} = -183\ kJ \cdot mol^{-1}, \tag{14}$$

$$3Fe + C \leftrightarrow Fe_3C,\ \Delta H^{\theta}_{298K} = 23\ kJ \cdot mol^{-1}, \tag{15}$$

The metallization degree, surface area, and porosity [80] influence the DRI carburization rate, along with the presence of active elements in the carburizing gas mixture and the operating temperature and pressure within the DR reactor [68]. The gas reactants are involved in both the reduction and carburization processes concurrently. Since iron ore pellets generally lack carbon-rich components, all the carbon in the DRI can be acquired during the reduction stage. However, this does not apply when H_2 is used as a reductant. DRI carburization is unlikely to occur, which may pose several challenges to EAF melting.

Nonetheless, carbon in the DRI can be presented as bonded carbon: cementite Fe_3C (possible formation of an intermediate carbide Fe_4C before conversion into Fe_3C or Fe_5C_2), a free carbon (graphite), or a combination of both [35]. The carbon of Fe_3C is part of the iron structure, while graphite is not chemically bonded with iron and adheres to the pellet surface or the pore surface, as mentioned in a study by He and Pistorius [81]. However, if this is an issue in the case of DRI, then in the case of using HBI, the graphite present on the surface will be compacted into a dense briquette with a low specific surface area.

The carbon contained in DRI produced, for instance, via a shaft furnace, is usually ≥90% Fe_3C [58]. Fe_3C can decompose at any temperature within the range of 500–900 °C [82]. However, the Fe_3C phase maintains its highest stability at temperatures ranging from 730 to 750 °C. Its decomposition rate accelerates at a temperature decrease from 750 °C and a temperature increase from 770 °C (Figure 2).

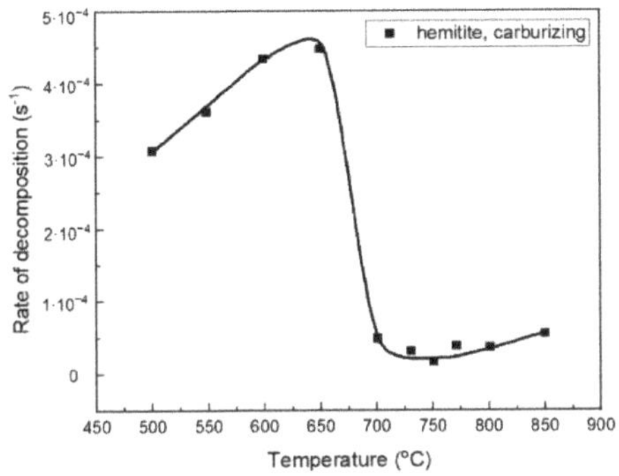

Figure 2. The rate of Fe_3C decomposition depends on temperature, according to a study by Longbottom et al., and is adapted with permission from Ref. [82]. Copyright 2007, Springer Nature.

According to Kumar et al. [83], CO is the main carburization gas, and applying it at 600 °C allows a faster processing rate. Using only CO, or CO with a small amount of H_2, leads to carbon deposition on the surface, particularly at high temperatures, which is undesirable. This is consistent with findings by Olsson et al. [84], who observed that the rate of carbon deposition can increase by up to 70% with H_2 concentration; beyond this point, a higher percentage of H_2 results in a decreased effect. However, adding H_2 to the gas mixture enhances the carburization rate. Ali et al. [22] reported that concentration in the reducing gas mixture enhances the growth of the Fe_3C and carbon deposition up to a certain degree. The highest Fe_3C formation was determined with an H_2:CO ratio of 25:75 at both temperatures of 973 K (~700 °C) and 1059 K (~786 °C), as shown in Figure 3.

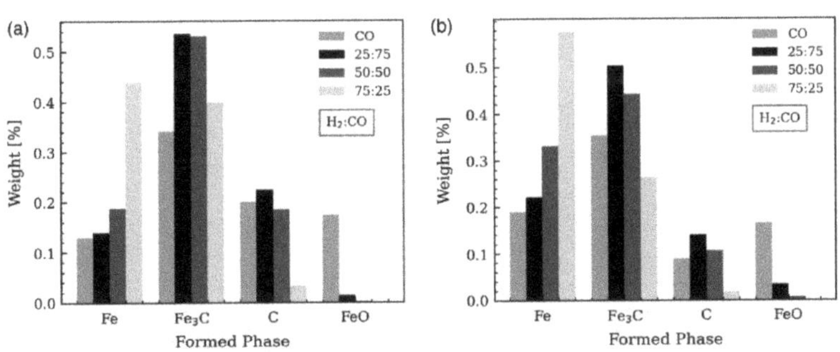

Figure 3. Different phases in the 55% reduced pellets with different gas compositions at: (**a**) 973 K (~700 °C) and (**b**) 1059 K (~786 °C), reprinted from Ref. [22].

In turn, CO_2 reduces the rate at which Fe_3C forms, so its presence in the carburizing gas mixture should be kept to a minimum. Furthermore, gas mixtures primarily composed of CH_4 achieve the fastest carburization rates at temperatures higher than 800 °C. The carburization process and its mechanism are detailed in [85–88].

On the one hand, charging Fe_3C-rich pellets to the EAF compared to graphite-rich pellets can be favorable. The heat formation of Fe_3C is positive, which means that melting of Fe_3C will require lower EAF energy compared to melting a mixture of iron and graphite [68]. Contrary to the abovementioned conclusion, in studies [89,90], it was reported that the DRI pellets, whether they contain graphite or Fe_3C and regardless of the carbon content, show similar melting enthalpies at 1600 °C, melting durations, and initial melting temperatures. This initial melting temperature is approximately 1147 °C, which is near the eutectic temperature for both Fe-Fe_3C and Fe-C systems. A notable difference was found in the melting behavior between graphitic and carbide-containing. DRI samples with predominantly Fe_3C melted uniformly, while those with graphite carbon displayed localized, spherical liquid areas that slowly developed over time.

Additionally, combining iron with carbon as Fe_3C can associate with the retardation of DRI reoxidation [91]. This advantage can be explained by the lower affinity of Fe_3C for oxygen than metallic iron.

4. Passivation of DRI

4.1. Reoxidation Issue of DRI

In general, removing oxygen from iron ore in the solid state forms numerous microscopic pores, creating a sponge-like texture to the iron [92]. The porosity of the DRI [93] is influenced by the choice of raw materials and the operational conditions prevalent during the DR process. Additionally, variation in reoxidation can be attributed to the differing carbon contents, which the reducing agent determines when used [94]. Therefore, it is important to mention the main mechanisms of reoxidation and the existing methods for mitigating these effects.

Reoxidation can occur via two mechanisms [95]: oxidation in the presence of air atmosphere and corrosion in an aqueous water-based environment. The chemical reactions that lead to the reoxidation of DRI generally release heat, making them exothermic. However, the extent of these reactions can be inhibited due to the carburization phenomenon discussed previously, which offers benefit not only to the EAF but also in terms of hindering reoxidation, owing to the blocking pores [96] resulting from the reduction process and the protective barrier formed by Fe_3C, which shields the material from fast reoxidation.

Dry oxidation progresses gradually at cooler temperatures, whereas the interaction with aqueous water unfolds swiftly and is often the catalyst for higher temperatures to a point where oxidation reactions begin at accelerated rates and turn self-sustaining, caused by rust layers [95]. Initially, reoxidation takes place at lower temperatures due to the exposure of DRI to air, factoring in moisture, according to Equations (16)–(21):

$$2Fe + 2H_2O + O_2 \leftrightarrow 2Fe(OH)_2, \tag{16}$$

$$2Fe(OH)_2 + \frac{1}{2}O_2 \rightarrow 2\gamma\text{—}FeO(OH) + H_2O, \tag{17}$$

$$Fe(OH)_2 + 2\gamma\text{—}FeO(OH) \rightarrow Fe_3O_4 + H_2O, \tag{18}$$

$$2Fe + H_2O + \frac{3}{2}O_2 \leftrightarrow 2FeO(OH), \tag{19}$$

$$3Fe + 4H_2O \leftrightarrow Fe_3O_4 + 4H_2, \tag{20}$$

$$Fe + 2H_2O \leftrightarrow FeO(OH) + \frac{3}{2}H_2, \tag{21}$$

Reactions (22)–(24) are highly exothermic, releasing a level of heat sufficient to boost additional oxidation processes:

$$3Fe + 2O_2 \leftrightarrow Fe_3O_4, \tag{22}$$

$$2Fe_3O_4 + \frac{1}{2}O_2 \leftrightarrow 3Fe_2O_3, \tag{23}$$

$$Fe + \frac{1}{2}O_2 \leftrightarrow FeO, \tag{24}$$

Implementing an inertization [97] or passivation technique enhances the stability of DRI for secure handling, transport, and storage. Various technologies exist for DRI passivation, as shown in Figure 4, each presenting its own set of advantages and disadvantages. However, none match the relatively higher efficiency of hot briquetting.

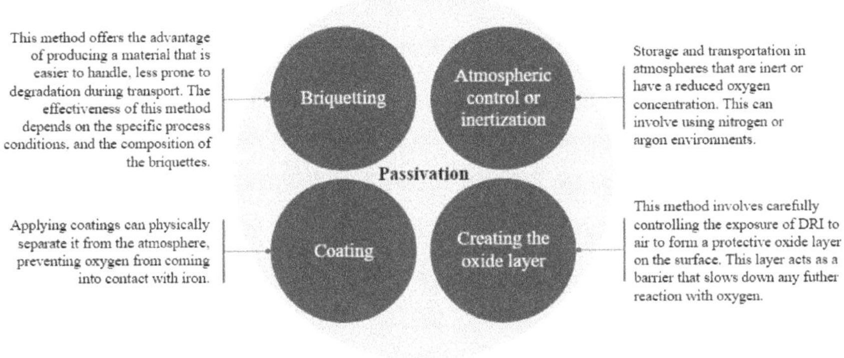

Figure 4. Main passivation techniques for DRI.

4.2. Creating the Oxide Layer

DRI can be maintained in an environment containing minimal oxygen to prevent reoxidation and create a thin oxide layer over all exposed surfaces. The reoxidation rate of DRI depends significantly on the morphological structure [98]. The morphological structure of DRI depends on the chemical composition of the iron ore, physicochemical properties of the pellets, reduction temperature, and gas compositions. If the time interval between reduction and oxidation is short, the reoxidation rate of DRI under air can be high. According to studies [95,99], the reoxidation process of iron initially does not maintain a constant temperature due to its exothermic nature, which heats the iron further. This process unfolds in three stages: the first phase shows quick oxidation, and the surface chemical reactions dominate; the second phase sees a gradual slowdown in oxidation, and the process indicates internal diffusion; the third phase has a low and stable oxidation rate, limited by slow diffusion, with a potential slight increase if the temperature rises. In conclusion, diffusion within the iron and forming cavities at the interface between the oxide layer and iron grains become the key mechanisms.

Abd Elmomen [94] oxidized reduced single DRI pellets in ambient air at an average temperature of 27 °C. The reaction duration varied from 1 to 50 days. The degree of reoxidation was determined by weight gain over time and the mass balance method. The reoxidation process proceeded topochemical and was controlled by the apparent rate of chemical reaction at the interface between the oxide layer and the unreacted core of metallic iron. It was concluded that the increase in the reoxidation degree and the accompanying decrease in the metallization degree depended on the source of oxide pellets used for DRI production.

In another study by Abd Elmomen [100], the kinetics of DRI reoxidation under stagnant air at temperatures between 150 and 450 °C were studied using two series of experiments. The first is in the relatively low-temperature range between 150 and 250 °C, and the second is at higher temperatures from 300 to 450 °C. In the mentioned low-temperature range, the reoxidation process was governed by a chemical reaction at the gas-unreacted core. No significant reoxidation occurred below 200 °C. In the relatively higher temperature range (300–450 °C), the reoxidation process proved to be controlled by pore diffusion.

Upadhya [101] also reported that below 220 °C, no significant reoxidation was observed. Additionally, it was found that the reoxidation rate became independent of the airflow above a critical flow rate of 0.85 $L \cdot min^{-1}$, indicating that at this stage, the influence of the gas boundary layer around the DRI pellet becomes negligible. The auto-ignition temperature of these DRI pellets was within the range of 200–230 °C.

Medina [102] reported that the reoxidation rate under dry air increased with increasing test temperatures to a maximum of 400–500 °C. In contrast, further increases in the test temperature reduced the rate of DRI reoxidation.

For pellets, after reduction in a hydrogen atmosphere, in the study by Cavaliere et al. [95], a very slow weight increase for temperatures up to 500 °C under air due to reoxidation was observed. Further, reoxidation occurred more rapidly at higher temperatures, with a weight increase of 4% after 10 min of exposure to 700 °C under air.

Moreover, there is a difference in reoxidation behavior between gas-based and coal-based DRI. In a moist oxygen-containing atmosphere, the activation energy for high-carbon gas-based DRI can be around 55 kJ·mol^{-1} (non-isothermal, heating rate of 15 °C·min^{-1}), and activation energy for coal-based DRI can be around 20–35 kJ·mol^{-1} (non-isothermal, heating rate of 15 °C·min^{-1}) [103,104]. Moreover, Bandopadhyay et al. [103] concluded that the gas-based DRI oxidizes relatively more quickly at approximately 447 °C, but at temperatures above 527 °C, the oxidation rate slows down, presumably because of the simultaneous oxidation of carbon.

4.3. *Hot Briquetted Iron*

Combining a high specific surface area and a porous structure in DRI increases susceptibility to reoxidation reactions [105]. The HBI was created in response to the issues with the shipping of DRI noted above, along with the need to briquette the fine metallic iron produced in the fluidized bed reduction processes [58] and pellets or lump ore from shaft furnace processes [16]. HBI is the compressed and densified version of DRI [106]. The material should be at a temperature of 650–700 °C before being compressed between two rollers. This process forms briquettes that typically measure between 90 and 140 mm in length, 48–58 mm in width, and 20–50 mm in thickness [107,108] and show a bulk density of 2.4–3.3 t·m^{-3} and an apparent density of 5.0–5.5 t·m^{-3} (Table 5). The variation in chemical and physical properties between DRI and HBI branches from the production pathways they undergo. Factors such as the grade and blend of the used ore, the nature of gangue materials, metallization degree, the use of additives, and the briquetting conditions also contribute to this variation.

Table 5. HBI physical properties, adapted from Refs. [35,58,64,77,78,108].

Characteristics	HBI
Bulk density, t·m^{-3}	2.4–3.3
Apparent density, t·m^{-3}	5.0–5.5
Specific surface area, m^2·g^{-1}	~0.75
Volumetric porosity, %	~21.0
Weight, g	500.0–700.0
Water absorption (saturated), %	~3.0
Fines (−4 mm), %	1.0–3.0

Compared to DRI, HBI has higher density and lower porosity. Due to its lower porosity, HBI tends to absorb less environmental moisture. In addition, HBI exhibits better wear and corrosion resistance due to its high bulk density, decreasing reactivity by one to two orders of magnitude [51,109].

HBI may undergo partial oxidation during handling, transportation, and storage processes, reducing its metallization degree. Environmental factors, along with production and briquetting parameters like the type of ore being reduced, as well as the pressure and temperature used during briquetting, significantly impact the rate of reoxidation. These variables influence the density of the briquettes, thereby affecting their reoxidation behavior, which can lead to the creation of various phases (Figure 5), for instance, ferrous hydroxide ($Fe(OH)_2$) [110], ferrous oxide (FeO), etc. Additionally, corrosion products can be detected as lepidocrocite (γ-FeO(OH)), iron trihydroxide ($Fe(OH)_3$), magnetite (Fe_3O_4), goethite (γ-FeO(OH)), maghemite (γ-Fe_2O_3), etc. [111].

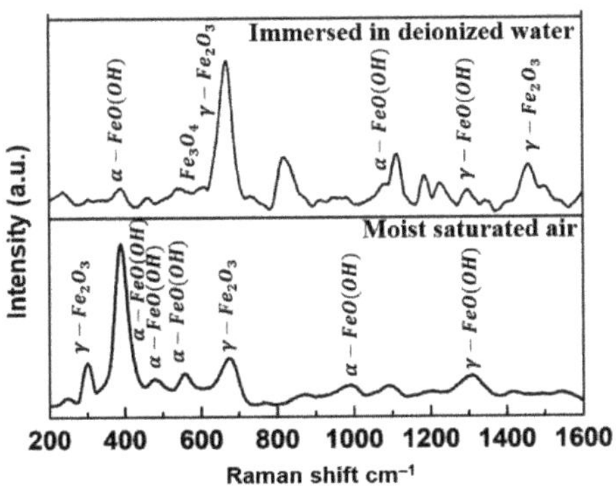

Figure 5. Raman spectra of possible phases can be detected in reoxidized HBI under different conditions, adapted from Ref. [109].

In this regard, Daghagheleh et al. [109] tested HBI samples under five different prepared climatic conditions for four months. Depending on the condition, different oxidation behaviors were observed. The results are shown in Figure 6 for both wet and dried HBI samples and the calculated metallization loss based on chemical analysis. HBI samples, which were exposed to ambient air conditions, developed nearly no reoxidation. The reoxidation was raised when the HBI samples were exposed to moisture-saturated air conditions. The reoxidation was further accelerated if the HBI samples were immersed constantly in deionized water. Immersing in process water has accelerated the reoxidation further. The worst case was experienced by samples periodically immersed in process water twice a week, each time for 1h comparison. The final metallization loss values were 0.92, 3.24, 4.26, 5.57, and 6.96 wt.% under ambient air, moist saturated air, immersed in deionized water, immersed in process water, and periodically immersed in process water, respectively. The porosity of the most reoxidized HBI samples was smaller than the raw HBI. The closure of cracks and pores by reoxidation products resulted in low porosity. The physical strength of the samples showed no significant difference before and after the reoxidation experiment. It can be concluded that interval and cyclic wetting and drying of HBI leads to the worst metallization loss problem. This case should be avoided from handling, storage, and shipment aspects.

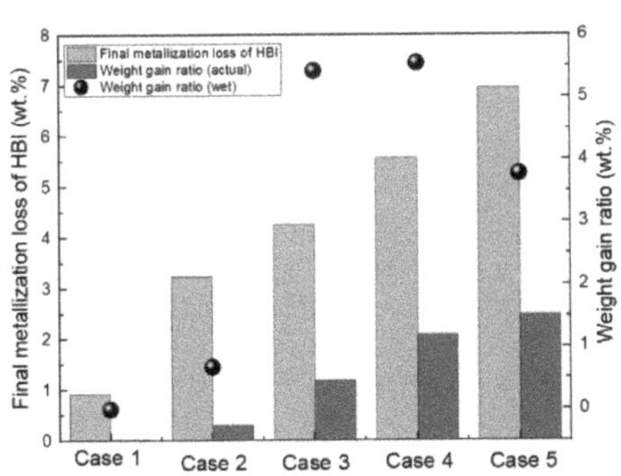

Figure 6. The final metallization loss of the dried HBI and weight gain ratio before and after drying were plotted based on data from Ref. [109].

The conditions under which the DRI was produced significantly influenced the reoxidation resistance of briquettes when exposed to oxidizing conditions [112,113].

From a laboratory-scale perspective, Gray et al. [114] reported that HBI derived from DRI with a 1.5 wt.% carbon content (produced at reducing temperatures between 500 and 800 °C, using a reducing gas composition of 70–75% H_2 and CO, along with H_2O, CO_2, CH_4, and N_2) showed more corrosion resistance compared to HBI from DRI with a 0.053 wt.% carbon content (produced at a reducing temperature of 630 °C with 100% H_2). This difference in performance is attributed to the variations in chemical compositions resulting from different reducing gas mixtures and ore types. Additionally, it was noted that briquettes with higher density (produced at 600 °C with a density of 5.2 $g \cdot cm^{-3}$) demonstrated better corrosion resistance than those with lower density (produced at 500 °C with a density of 4.2 $g \cdot cm^{-3}$).

Practically, there is a notable gap in understanding the reoxidation process, particularly with HBI, and measurements that should be considered subsequently for handling and transportation. In this context, the currently running HBI-C-Flex project is funded within the frame of the Research Fund for Coal and Steel (RFCS, Grant Agreement no. 101112479) [115] gains significance as it aims to test the reoxidation of HBI and DRI (with different carbon contents and low-grade iron ore application) under varying environmental conditions (including both dry and wet atmospheres and higher temperatures) to measure the exothermic reactions that occur during storage and transportation. Insights into the reactivity and stability of HBI, including dust formation, will be crucial for steel manufacturers. Further investigation will aid in improving logistics and enhancing knowledge to prevent dangerous situations in operations.

4.4. Other Methods

Applying coatings to DRI [51], such as sodium silicate, limestone, cement, or waxes [116,117], is a potential strategy for retardation of reoxidation. However, the disadvantages of coatings include the risk of additional contamination, increased costs, and the possibility of the coating being scratched during handling and transportation. Another disadvantage that can be considered is that coated DRI pellets can consume more electricity and graphite electrodes in the EAF compared to uncoated DRI pellets [118]. Nevertheless, to evaluate the effectiveness of the waxing process in preventing reoxidation of DRI, as detailed in [119], the experiment was conducted where the DRI was immersed in a bath of melted paraffin at temperatures between 110 and 120 °C for 3–5 s. It was observed that no heat was produced from the treated DRI after it was sprayed with a 5% sodium chloride solution, suggesting that corrosion did not occur. Additionally, there was no change in weight after the DRI was heated to 150 °C for 30 min in an air jet oven, indicating the absence of reoxidation and no loss of wax. Furthermore, no dust was released from the treated DRI samples following a tumbler index test.

After ammonia DR [120–124], it was noticed that the nitride passivates the otherwise highly active reduced iron, offering a safety-critical benefit for handling and logistics. Nitride formation presents a benefit of the ammonia DR process, as it enhances the aqueous corrosion resistance of iron. Specifically, during the quenching process (at temperatures below 700 °C), metallic iron within the sample undergoes nitriding in the presence of NH_3. Conversely, in the absence of NH_3 at 700 °C, nitrides decompose back into metallic iron. This indicates that the quenching conditions are crucial in determining the phases produced. Consequently, the phases present in the reduced ore can be effectively managed during the post-reduction stage [120].

Several factors, including the origin and type of iron ore, affect the stability, reactivity, and reducibility of DRI. Improving the resistance of DRI to reoxidation and spontaneous ignition involves prolonging the reduction process and increasing its temperature. The gas composition during DR impacts the quality of reduction and carbide stability, while the porosity of DRI increases its reoxidation occurrence. As mentioned above, briquetting DRI into HBI reduces its specific surface area and exposure to air, enhancing stability by forming a protective layer against further reoxidation.

5. Storage, Handling, and Transportation

Some main measurements can be extracted from [35,125–127] and generalized as follows. For safe DRI storage, it is critical to avoid storing any DRI hotter than 65 °C in bins or silos, instead keeping it isolated and not stacked higher than one meter. Safety measures include conducting regular gas analyses to keep oxygen below 3% and detecting hydrogen production. Monitoring and managing temperatures within storage units are essential, with procedures to seal and ventilate units if temperatures reach 65–75 °C. During extended storage, introducing inert gas prevents reoxidation, and ensuring slide gates remain closed maintains a stable, safe environment.

Additionally, according to [128], DRI can be categorized into three types: less reactive (category A), namely HBI; highly reactive (category B); and by-product fines (category C), which are generated after production and handling of DRI or HBI. Depending on the category, particular recommendations for loading and carriage are described in detail [127].

6. Application of DRI/HBI in an EAF Steelmaking

Process-wise, in DRI-based steelmaking, there are two primary methods for sourcing. The first method involves directly melting the DRI as soon as it exits the DRI production process. Alternatively, the second method utilizes HBI, which is cold, compacted, and treated for safe transportation. This approach offers flexibility by separating the DRI production process from steelmaking, allowing for more logistical and operational convenience [129]. HBI and hot DRI possess similar advantageous characteristics to those of cold DRI. However, for hot DRI, its temperature plays a crucial role in enhancing EAF efficiency while maintaining its high metallic content [130,131]. Hot DRI can be transported to an EAF at a temperature of ~700 °C [132] using various methods, including hot transport vessels, pneumatic transport systems, hot conveyors, and gravity systems [51,133]. This approach can lower electrical energy usage, leading to energy savings of approximately 20–30% [132] compared to traditional furnaces that are fed with cold materials. Additionally, according to Gonzalez et al. [134], as the initial temperature of the DRI increases, the melting rate augments; thus, tap-to-tap time decreases, promoting higher productivity. The full potential of this hot charging technology is not completely leveraged due to the differing operational requirements between EAF and DRI production processes [132]. Nevertheless, using hot DRI charging offers some benefits [134,135].

Typically, DRI/HBI can be charged with scrap in varying amounts to the EAF, influenced by local cost factors and material availability [19]. DRI/HBI can be within the range of 10–30% of the total charge [133] in the EAF. HBI, often added to the EAF alongside scrap using buckets, requires minimal adjustments to the existing EAF, mainly when HBI constitutes up to 10% of the charge. The design of the EAF can allow for an increased "hot heel" or the remaining melt volume, which can be up to 30% [136] of the total melt volume. This adjustment aids in the effective melting of DRI. Consequently, to accommodate the continuous addition of materials, maintain prolonged periods of a flat bath, and manage the increased need for lime and calcined dolomite for slag formation, adjustments to the power programs of the EAF are necessary.

In EAF, DRI/HBI demonstrates different characteristics than the steel scrap melting. This variation can be based on DRI/HBI/scrap share, metallization degree [137], and due to gangue and carbon within the DRI/HBI [138]. Since the DR process operates in a solid state, the gangue from the DR pellets is not removed but instead transferred into the EAF along with the DRI. This results in increased amounts of slag during the steelmaking process due to the higher levels of gangue content [139].

As the DRI/scrap ratio increases and the DRI metallization degree decreases, the iron oxide content in slag can increase. An increase in the SiO_2 content in DRI demands a higher CaO addition to keep the basicity ratio, subsequently increasing energy demands [77]. According to Kirschen et al. [19], the average iron loss in EAF using DRI can be greater than in using scrap, and controlling FeO content proves more challenging in EAF with DRI.

The standard deviation of FeO ranged from 3.9 to 5.5 for scrap and widened from 4.1 to 9.6 when the DRI usage was >50% (Figure 7).

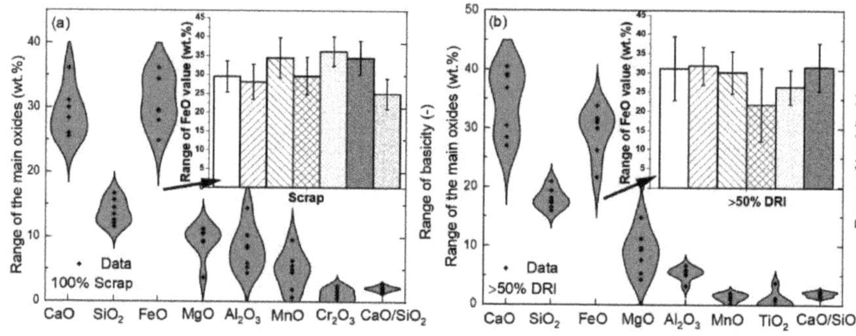

Figure 7. Range of main oxides in EAF slag depending on charged material: (**a**) 100% scrap; (**b**) >50% DRI, plotted based on data from Ref. [19].

Additionally, the yield from DRI inputs can be typically lower, which is attributed to the oxide gangue content in DRI. This contrasts with the yields from melting 100% scrap, which range from 90 to 94% (Table 6). The melting time and tap-to-tap time for DRI-based production are considerably longer than those used for scrap [19].

Table 6. Production parameters of EAF charge with scrap and scrap/DRI for low alloyed steel grades, adapted from Refs. [16,19].

Production Parameters	100% Scrap	80–95% DRI
Share of DRI/HBI, % procedures	0–5 (HBI)	60–95 (DRI)
Electric energy demand, kWh·t^{-1}	310–460	530–680
Natural gas, $m^3 \cdot t^{-1}$	3–10	0–2
Oxygen, $m^3 \cdot t^{-1}$	25–40	20–35
Coal and carbon fines, kg·t^{-1}	2–9	8–17
Slag former, kg·t^{-1}	23–35	27–60
Tap temperature, °C	1600–1635	1600–1635
Tap-to-tap time, min	50–60	60–100
Metal yield, %	90–94	87–92

Concerning carbon in DRI, it serves several reasons [10,79]: first, carbon is essential for the complete metallization of iron; second, it serves as an additional energy source. When oxygen is injected to burn the carbon, it reduces the need for electricity, thereby accelerating the melting process of the materials charged. Third, carbon is necessary to create a foamy slag in the EAF, benefiting the process (Figure 8). For instance, for each 1.0 wt.% of iron in the form of FeO within the DRI, theoretically, 0.215 wt.% of carbon is needed to reduce it through the endothermic reaction $FeO + C \rightarrow Fe + CO$ [35]. Furthermore, carbon can be engaged in various oxidation reactions with CO formation, such as $CO_2 + C \rightarrow 2CO$ ($\Delta H \geq 0$) and $C + \frac{1}{2}O_2 \rightarrow CO$ ($\Delta H \leq 0$), which can facilitate stirring. Different from carbon materials that are charged or injected [140–142], DRI contains carbon but is free from volatile matter, ash, and sulfur, which can affect the melting process and/or the quality of steel.

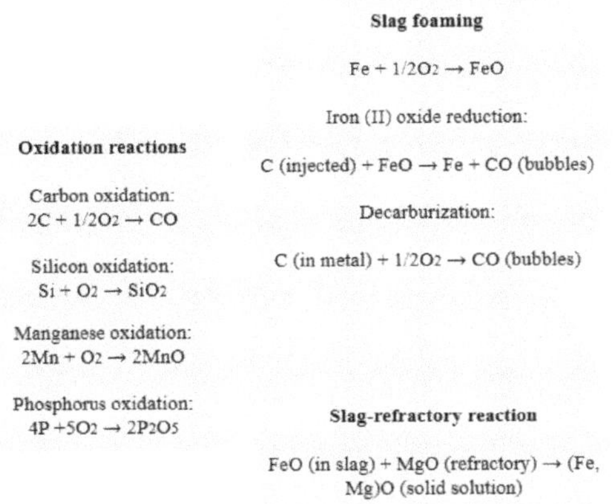

Figure 8. Chemical reactions in an electric arc furnace, adapted from Ref. [143].

According to Memoli et al. [144] DRI is more effective than charged carbon material in the bath reactions, yields, and purity of the carbon in the steel.

Lule et al. [145] highlighted results from the ArcelorMittal Lázaro Cardenas steelmaking shop, focusing on N_2 behavior. The use of high-carbon and a high proportion of DRI proved advantageous for producing N_2-critical steel grades, attributed to the significant evolution of CO bubbles allowing the N_2 transfer from the liquid steel to the gas atmosphere in the bubble. The same behavior was also claimed by Memoli et al. [144].

Several studies have also been devoted to investigating the influence of the DRI amount on the steelmaking process parameters. With increasing the percentage of DRI in the charge, there can be a noticeable decrease in metallic yield, tramp elements, and sulfur, as reported in [146]. Similar observations of improvement in steel quality by increasing the DRI percentage (varying range of DRI of 0–90%) have been noticed by Elkader et al. [66]. The reduction in tramp (copper, tin, nickel, and chromium), phosphorus, sulfur, and nitrogen were correlated with increased DRI proportions. Conversely, a rise in the DRI content resulted in greater per-ton consumption of electric power, oxygen, coke, and fluxing agents required to produce liquid steel. Additionally, as the percentage of DRI in the charge increased, there was an extension in both power-on and overall tap-to-tap times.

Hassan et al. [147] studied the impact of DRI/HBI on the steelmaking process parameters (Figure 9a,b), along with methods and proportions (case 1: 100% steel scrap; case 2: 100% HBI in the proportion of metallic charge of 17.36%; case 3: 100% HBI in the proportion of metallic charge of 26.76%; case 4: 50% HBI: 50% DRI in the proportion of metallic charge of 45.16%; case 5: 100% DRI in the proportion of metallic charge of 75.02%; and case 6: 100% DRI in the proportion of metallic charge of 100% of DRI) under the temperature condition of 1640 ± 10 °C. Increasing the amount of lower-quality DRI and/or HBI to 45% of the metallic charge decreased steel yield to 85.32%. Conversely, the yield increased to 87.57% when the input consisted of 75–100% high-quality DRI, continuously fed. This led to a 21-min decrease in tap-to-tap time and a 7% decrease in electrode usage. Increasing the proportion of lower-quality DRI in the EAF from 45% to 100% resulted in a 4 tons/heat decrease in liquid steel and caused lime usage to increase from 4 to 6 $kg \cdot t_{LS}^{-1}$.

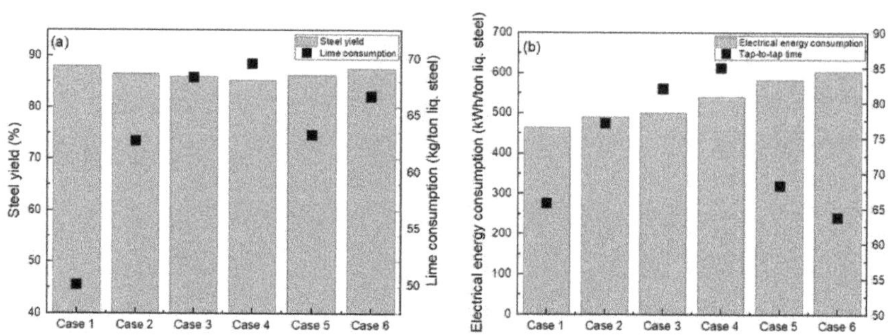

Figure 9. Impact of DRI/HBI on the selected steelmaking process parameters: (**a**) steel yield and lime consumption; (**b**) electrical energy consumption and tap-to-tap time, plotted based on data from Ref. [147].

The use of DRI can affect refractory degradation by altering the composition of slag, specifically reducing the CaO/SiO_2 basicity [148]. Heo and Park [149] investigated the interfacial reactions between the EAF slag of 20 wt.% CaO—16.9 wt.% SiO_2—39.6 wt.% FeO—3.2 wt.% MgO—13.1 wt.% Al_2O_3—7.2 wt.% MnO and magnesia refractory, varying the addition of DRI (0, 10, 20, 30 wt.%). Their investigations were conducted at 1550 °C under an argon atmosphere. They suggested that controlling the amount of DRI is essential to reduce refractory degradation during the EAF process, as increasing DRI content can decrease the CaO/SiO_2. Further, Heo and Park [150] conducted research on how temperature variations between 1550 and 1650 °C impact the chemical interaction at the boundary between the slag, composed of CaO—SiO_2—MgO—35 wt.% FeO—10 wt.% Al_2O_3—5 wt.% MnO, and the refractory under conditions incorporating 20 wt.% DRI in the EAF. Their findings suggest that precise temperature control in an EAF process that utilizes DRI plays a more crucial role in reducing refractory wear than the proportion of DRI used.

Furthermore, DRI melting in the EAF process encompasses heat, mass, and momentum transfer, as well as a multiphase system comprising liquid slag, liquid steel, evolving gases, and solid particles [134]. As mentioned above, the physical properties and chemical composition of DR influence the melting process. Factors such as the method of charging, furnace type, bath temperature, the chemical composition of the molten phases, and the circulation of fluids within the furnace and around the particles should also be considered.

Understanding how DRI behaves when immersed in liquid slag is based on several research data obtained under a laboratory scale, which will be discussed further and tends to be more about qualitative or visual descriptions of experiments rather than quantitative results. Based on visual observations of the DRI behavior in liquid slag, it can be categorized as follows: sink through, float inside the slag, or remain on top before complete decarburization.

The study by Li and Barati [132] is one of the most thorough in explaining the behavior of DRI in slag. To understand the decarburization and melting behavior of DRI pellets, the SiO_2-Al_2O_3-CaO-MgO slag with various FeO contents (10–25 wt.%) and basicity $(MgO + CaO)/(SiO_2 + Al_2O_3)$, ranging from 1.5 to 2.5, has been used. The result showed that the decarburization of DRI in slag at 1600 °C takes place in two stages: initially, the reaction is between FeO and the carbon within the pellet, driven by heat transfer from the slag to the pellet. The subsequent phase involves FeO from the slag reacting with the remaining carbon in the DRI. The reaction rate depends on how FeO moves within the slag and is heavily influenced by its concentration. Additionally, in the study, three possible behaviors of DRI were observed in slag. In the first case, the DRI pellet immediately sinks to the bottom of the crucible, creating a thin and uneven gas layer that occasionally moves the pellet. This was attributed to the weaker gas evolution of $8.35 \cdot 10^{-5}$ mol·s^{-1} in the first stage of decarburization. In the second case, the DRI pellet stays buoyant and moves within the slag due to a thicker gas halo forming around it until it turns into a liquid droplet and sinks. In the third case, high slag viscosity causes the pellet to partially immerse and restrict its movement, linked to a higher gas-evolution rate of $2.11 \cdot 10^{-4}$ mol·s^{-1} in the first stage of decarburization. In all cases, it was observed that the pellet began to shrink after 16–26 s of immersion, and a dense shell was formed around the pellet.

It should be mentioned that firstly, Mulholland et al. [151] observed the gas layer in their research on the reaction between iron-carbon-sulfur drops (0.8–4.5 wt.% C, 0.04–1.0 wt.% S) and steelmaking slag (47 wt.% CaO—38 wt.% SiO_2—15 wt.% Al_2O_3—10–30 wt.% Fe_2O_3) via X-ray technique at 1723–1873 K (approximately at 1450–1600 °C). It was concluded that the Fe-C-S drop into the slag initially floats, followed by slag foaming. Subsequently, the drop sank, and the foamy slag layer disappeared. Notably, it was impossible to determine the different velocities of the drop's movement to the bottom of the crucible, which varied widely, ranging within 1–36 cm·s^{-1}. According to Min and Fruehan [152], the reaction apparently occurs in two steps: one at the gas slag interface and the other at the gas metal interface, with CO_2 diffusion through the gas halo.

A similar manner was reported by Goldstein et al. [153] when studying the behavior of cold DRI pellets dropped into a slag system of 40 wt.% CaO—41 wt.% SiO_2—10 wt.% Al_2O_3—9 wt.% FeO at 1460 °C. The gas rate produced from the FeO reduction in the DRI was measured by a constant volume pressure increase method. It indicated a short incubation period of approximately 2–4 s, followed by rapid gas evolution from the pellet that lasted between 20 and 30 s. The gas halo formed around the pellet, causing the pellet to remain buoyant in the slag.

From the point of view of the dependence of melting time on several physical properties of DRI, Gonzalez et al. [134] suggested that the melting time is significantly affected by the thermal conductivity of DRI. They noted that increasing the metallization degree of DRI and reducing its porosity can raise thermal conductivity. Additionally, it was highlighted that an increase in porosity adversely affects melting time, leading to longer durations for the melting process.

Sharifi et al. [154] investigated how the initial carbon content and the preheating temperature of DRI affect the reaction rate, aiming for a slag composition of 22 wt.% FeO—42 wt.% CaO—22 wt.% SiO_2—4 wt.% Al_2O_3—10 wt.% MgO, with temperatures ranging from 1500 to 1600 °C. They determined that the decarburization of DRI could occur in two distinct stages. Furthermore, preheating DRI pellets was found to accelerate the initiation of the reaction between carbon and FeO, thus shortening the total reaction time (Figure 10). This reduction suggests a potential for enhancing the productivity of DRI-based steel production through the preheating of DRI.

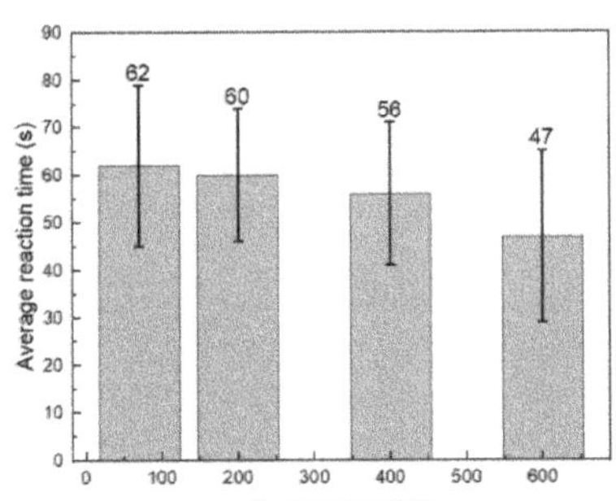

Figure 10. Average reaction times for four preheating temperatures of DRI, plotted based on data from Ref. [154].

Considering the conclusion of the study by Sharifi et al. [154] that the decarburization of DRI pellets takes place in two stages: a reaction between FeO and carbon inside the pellet, followed by a reaction of the remaining carbon with the FeO inside the slag, Kiasaraei [155] reported that the duration of each stage varies based on the carbon and FeO levels in the

DRI, with the second stage potentially being negligible in cases of low carbon content. The study highlights the significant impact of carbon content on decarburization rates, noting an increase in reaction rate for both phases with higher carbon levels, especially in the second stage. Additionally, it was determined that heat transfer mechanisms predominantly control the reaction rate in the first stage, indicating its role as an important factor. Li and Barati came to the same conclusion [132].

Furthermore, research was conducted on how the size of DRI pellets influences the process, and different conclusions were reported [153,156]. Sadrnezhaad [157] stated that the optimum size of the DRI pellets was around 7 mm for experiments. The optimum feeding rate for the lowest energy consumption was 12.5 $g{\cdot}s^{-1}$ for continuous feeding of DRI for a 25 kg induction furnace at 1550 °C. In addition, it was determined that a DRI pellet with a metallization degree of 92.5% and carbon content of 1.6 wt.% could be assumed to behave like an isolated gassing particle floating on the surface of the liquid melt. The heating and melting process took a few seconds compared to the case with the highly metalized DRI of 97%.

A notable reduction in melting rate can be observed for DRI particles with large sizes, for instance, above 10 mm, according to González et al. [134], and when the arc length is elevated [21]. However, with an increase in the initial particle size, the thickness of the frozen shell, which plays a large role in the melting rate of DRI, can rise. In addition, extended arc operation can harm the refractory of a furnace by creating intense hot spots and reducing thermal efficiency through uncontrolled radiation. This issue can be mitigated by employing foamy slag.

The productivity of an EAF that continuously melts DRI depends on the chemical composition of DRI, the feed rate of the DRI, and effective slag foaming. The yield of liquid steel when using DRI or HBI depends on the metallization degree, the total content of gangue, and carbon injection. In general, achieving a higher yield of liquid steel requires a greater degree of metallization in DRI or HBI.

7. Conclusions and Prospects

The DR-EAF route presents an alternative to the conventional BF-BOF route. The production route of DRI can be divided into one that uses gas reductants (converted NG, syngas, H_2, etc.) and another that relies on coal. Investigating various H_2-containing reducing gases and their combinations can be relevant. In this regard, exploring alternative H_2 carriers like ammonia is viable, along with the potential to not only be a source of the required H_2 but also provide a solution against reoxidation.

In the case of DRI carburization, when carbon or carbon-containing gases (CO/CH_4) are applied to facilitate effective carburization, it is important that the gas mixture used minimizes CO_2 and optimizes the ratio of CO to H_2. The presence of CH_4 in the gas mixture can also accelerate the carburization process at temperatures above 800 °C. Alongside this, the presence of carbon in DRI from the perspective of EAF can have some benefits than charged/injected carbon material.

Concerning reoxidation behavior, and based on the research results, several general conclusions can be drawn. Significant reoxidation is usually not observed at low temperatures under air. However, in some cases, further reoxidation can occur more rapidly at higher temperatures, up to 700 °C under air conditions. The reoxidation process can be accompanied by a degree of loss of metallization. As mentioned earlier, the briquetting of DRI is one of the most efficient methods to prevent reoxidation. However, the pre-conditions and environmental factors can influence the reoxidation behavior of HBI. Further research on the reoxidation behavior of HBI could enhance understanding of its reoxidation mechanisms and enable the development of more effective mitigation strategies for handling and transportation. This is particularly important as most existing studies have primarily concentrated on the reoxidation of DRI rather than HBI. The analysis presented did not fully explore the reoxidation behavior of DRI/HBI produced using H_2 as a reducing agent due to a scarcity of such findings in the existing literature. This gap highlights the potential value of research in this area, particularly regarding the establishment of handling and transportation conditions.

Considering the possibility of further research prospects in terms of processing low-quality iron ore with high gangue content, the DR-EAF route is advantageous for processing high-quality iron ore (≥67% Fe), allowing flexible supplementation of DRI and HBI with scrap based on market conditions. However, the availability of high-grade ores is limited. In contrast, from the point of view of prospects, the DR-electric smelting furnace-BOF route is more suitable for processing low-quality iron ore with high gangue content. This route involves further refining the pre-melt in a conventional BOF and offers the additional benefit of using smelter slag in the cement industry.

Author Contributions: Conceptualization, L.K., S.L. and O.D.; methodology, L.K., S.L. and O.D.; resources, J.R.; data curation, L.K., S.L. and O.D.; writing—original draft preparation, L.K.; writing—review and editing, L.K., S.L., J.R., M.L., L.S. and O.D.; project administration, L.S. and M.L. All authors have read and agreed to the published version of the manuscript.

Funding: This research receives funding from the European Union (RFCS Research Fund for Coal and Steel) within the frame of the "HBI C-Flex" project, Grant Agreement number 101112479.

Data Availability Statement: The raw data supporting the conclusions of this article will be made available by the authors on request.

Acknowledgments: The authors gratefully acknowledge the funding support of K1-MET GmbH, a metallurgical competence center. The research program of the K1-MET competence center is supported by COMET (Competence Center for Excellent Technologies), the Austrian program for competence centers. COMET is funded by the Federal Ministry for Climate Action, Environment, Energy, Mobility, Innovation, and Technology, the Federal Ministry for Labour and Economy, the Federal States of Upper Austria, Tyrol, and Styria, as well as the Styrian Business Promotion Agency (SFG) and the Standortagentur Tyrol. Furthermore, Upper Austrian Research GmbH continuously supports K1-MET. Besides the public funding from COMET, this research project is partially financed by scientific and industrial partners.

Conflicts of Interest: Authors Lina Kieush, Stefanie Lesiak, Johannes Rieger, Melanie Leitner, Lukas Schmidt were employed by the company K1-MET GmbH. The remaining authors declare that the research was conducted in the absence of any commercial or financial relationships that could be construed as a potential conflict of interest.

References

1. Dall'Osto, G.; Mombelli, D.; Mapelli, C. Consequences of the Direct Reduction and Electric Steelmaking Grid Creation on the Italian Steel Sector. *Metals* **2024**, *14*, 311. [CrossRef]
2. Pfeiffer, A. Evaluation of the Smelting Behaviour of Direct Reduced Iron. Ph.D. thesis, Montanuniversitaet Leoben, Leoben, Austria, 2023.
3. Béchara, R.; Hamadeh, H.; Mirgaux, O.; Patisson, F. Optimization of the Iron Ore Direct Reduction Process through Multiscale Process Modeling. *Materials* **2018**, *11*, 1094. [CrossRef]
4. Fei, Y.; Guan, X.; Kuang, S.; Yu, A.; Yang, N. A Review on the Modeling and Simulation of Shaft Furnace Hydrogen Metallurgy: A Chemical Engineering Perspective. *ACS Eng. Au* **2023**, *4*, 145–165. [CrossRef]
5. Diez, M.A.; Centeno, T.A.; Amado-Fierro, Á. Coal Use for Iron and Steel Production in Low-Carbon Transition Scenarios. In *The Coal Handbook*; Elsevier: Amsterdam, The Netherlands, 2023; pp. 493–546, ISBN 978-0-12-824327-5.
6. Hammam, A.; Cao, Y.; El-Geassy, A.-H.A.; El-Sadek, M.H.; Li, Y.; Wei, H.; Omran, M.; Yu, Y. Non-Isothermal Reduction Kinetics of Iron Ore Fines with Carbon-Bearing Materials. *Metals* **2021**, *11*, 1137. [CrossRef]
7. Boretti, A. The Perspective of Hydrogen Direct Reduction of Iron. *J. Clean. Prod.* **2023**, *429*, 139585. [CrossRef]
8. Devlin, A.; Kossen, J.; Goldie-Jones, H.; Yang, A. Global Green Hydrogen-Based Steel Opportunities Surrounding High Quality Renewable Energy and Iron Ore Deposits. *Nat. Commun.* **2023**, *14*, 2578. [CrossRef]
9. Ling, J.; Yang, H.; Tian, G.; Cheng, J.; Wang, X.; Yu, X. Direct Reduction of Iron to Facilitate Net Zero Emissions in the Steel Industry: A Review of Research Progress at Different Scales. *J. Clean. Prod.* **2024**, *441*, 140933. [CrossRef]
10. Rechberger, K.; Spanlang, A.; Sasiain Conde, A.; Wolfmeir, H.; Harris, C. Green Hydrogen-Based Direct Reduction for Low-Carbon Steelmaking. *Steel Res. Int.* **2020**, *91*, 2000110. [CrossRef]
11. DRI Production. Available online: https://www.metallics.org/dri-production.html (accessed on 27 July 2024).
12. 2022 World Direct Reduction Statistics. Available online: https://www.midrex.com/wp-content/uploads/MidrexSTATSBook2022.pdf (accessed on 7 March 2024).

13. Guo, D.; Li, Y.; Cui, B.; Chen, Z.; Luo, S.; Xiao, B.; Zhu, H.; Hu, M. Direct Reduction of Iron Ore/Biomass Composite Pellets Using Simulated Biomass-Derived Syngas: Experimental Analysis and Kinetic Modelling. *Chem. Eng. J.* **2017**, *327*, 822–830. [CrossRef]
14. Xu, C.C.; Cang, D. A Brief Overview of Low CO_2 Emission Technologies for Iron and Steel Making. *J. Iron Steel Res. Int.* **2010**, *17*, 1–7. [CrossRef]
15. Ramakgala, C.; Danha, G. A Review of Ironmaking by Direct Reduction Processes: Quality Requirements and Sustainability. *Procedia Manuf.* **2019**, *35*, 242–245. [CrossRef]
16. Kirschen, M.; Badr, K.; Pfeifer, H. Influence of Direct Reduced Iron on the Energy Balance of the Electric Arc Furnace in Steel Industry. *Energy* **2011**, *36*, 6146–6155. [CrossRef]
17. Morales, R.D.; Conejo, A.N.; Rodriguez, H.H. Process Dynamics of Electric Arc Furnace during Direct Reduced Iron Melting. *Met. Mater. Trans. B* **2002**, *33*, 187–199. [CrossRef]
18. World Steel Association, Sustainability Indicators. Available online: https://worldsteel.org/steeltopics/sustainability/sustainability-indicators/ (accessed on 7 March 2024).
19. Kirschen, M.; Hay, T.; Echterhof, T. Process Improvements for Direct Reduced Iron Melting in the Electric Arc Furnace with Emphasis on Slag Operation. *Processes* **2021**, *9*, 402. [CrossRef]
20. Abolpour, B.; Afsahi, M.M.; Azizkarimi, M. Hydrogen Reduction of Magnetite Concentrate Particles. *Miner. Process. Extr. Metall.* **2021**, *130*, 59–72. [CrossRef]
21. Sanchez, J.L.G. Power Delivery from the Arc in AC Electric Arc Furnaces with Different Gas Atmospheres. *Steel Res. Int.* **2009**, *80*, 113–120. [CrossRef]
22. Ali, M.L.; Fradet, Q.; Riedel, U. Kinetic Mechanism Development for the Direct Reduction of Single Hematite Pellets in H_2/CO Atmospheres. *Steel Res. Int.* **2022**, *93*, 2200043. [CrossRef]
23. Ei-Geassy, A.A.; Shehata, K.A.; Ezz, S.Y. Mechanism of Iron Oxide Reduction with Hydrogen/Carbon Monoxide Mixtures. *ISIJ Int.* **1977**, *17*, 629–635. [CrossRef]
24. Scharm, C.; Küster, F.; Laabs, M.; Huang, Q.; Volkova, O.; Reinmöller, M.; Guhl, S.; Meyer, B. Direct Reduction of Iron Ore Pellets by H_2 and CO: In-Situ Investigation of the Structural Transformation and Reduction Progression Caused by Atmosphere and Temperature. *Miner. Eng.* **2022**, *180*, 107459. [CrossRef]
25. Sun, M.; Pang, K.; Barati, M.; Meng, X. Hydrogen-Based Reduction Technologies in Low-Carbon Sustainable Ironmaking and Steelmaking: A Review. *J. Sustain. Metall.* **2023**, *10*, 10–25. [CrossRef]
26. Liu, W.; Zuo, H.; Wang, J.; Xue, Q.; Ren, B.; Yang, F. The Production and Application of Hydrogen in Steel Industry. *Int. J. Hydrog. Energy* **2021**, *46*, 10548–10569. [CrossRef]
27. Abu Tahari, M.N.; Salleh, F.; Tengku Saharuddin, T.S.; Samsuri, A.; Samidin, S.; Yarmo, M.A. Influence of Hydrogen and Carbon Monoxide on Reduction Behavior of Iron Oxide at High Temperature: Effect on Reduction Gas Concentrations. *Int. J. Hydrog. Energy* **2021**, *46*, 24791–24805. [CrossRef]
28. Takenaka, Y.; Kimura, Y.; Narita, K.; Kaneko, D. Mathematical Model of Direct Reduction Shaft Furnace and Its Application to Actual Operations of a Model Plant. *Comput. Chem. Eng.* **1986**, *10*, 67–75. [CrossRef]
29. Hara, Y.; Sakawa, M.; Kondo, S. Mathematical Model of the Shaft Furnace for Reduction of Iron-Ore Pellet. *Tetsu-to-Hagane* **1976**, *62*, 315–323. [CrossRef] [PubMed]
30. Di, Z.; Li, Z.; Wei, R.; Liu, Y.; Meng, Q.; Chun, T.; Long, H.; Li, J.; Wang, P. Sticking Behaviour and Mechanism of Iron Ore Pellets in COREX Pre-Reduction Shaft Furnace. *Ironmak. Steelmak.* **2019**, *46*, 159–164. [CrossRef]
31. Heidari, A.; Niknahad, N.; Iljana, M.; Fabritius, T. A Review on the Kinetics of Iron Ore Reduction by Hydrogen. *Materials* **2021**, *14*, 7540. [CrossRef] [PubMed]
32. Souza Filho, I.R.; Springer, H.; Ma, Y.; Mahajan, A.; Da Silva, C.C.; Kulse, M.; Raabe, D. Green Steel at Its Crossroads: Hybrid Hydrogen-Based Reduction of Iron Ores. *J. Clean. Prod.* **2022**, *340*, 130805. [CrossRef]
33. Patisson, F.; Mirgaux, O. Hydrogen Ironmaking: How It Works. *Metals* **2020**, *10*, 922. [CrossRef]
34. Cavaliere, P.; Perrone, A.; Marsano, D. Effect of Reducing Atmosphere on the Direct Reduction of Iron Oxides Pellets. *Powder Technol.* **2023**, *426*, 118650. [CrossRef]
35. Anameric, B.; Kawatra, S.K. Properties and features of direct reduced iron. *Miner. Process. Extr. Metall. Rev.* **2007**, *28*, 59–116. [CrossRef]
36. Zhang, T.; Lei, C.; Zhu, Q. Reduction of Fine Iron Ore via a Two-Step Fluidized Bed Direct Reduction Process. *Powder Technol.* **2014**, *254*, 1–11. [CrossRef]
37. Wan, Z.; Huang, J.; Zhu, G.; Xu, Q. Numerical Simulation of the Operating Conditions for the Reduction of Iron Ore Powder in a Fluidized Bed Based on the CPFD Method. *Processes* **2022**, *10*, 1870. [CrossRef]
38. Mishra, S.; Roy, G.G. Effect of Amount of Carbon on the Reduction Efficiency of Iron Ore-Coal Composite Pellets in Multi-Layer Bed Rotary Hearth Furnace (RHF). *Met. Mater. Trans. B* **2016**, *47*, 2347–2356. [CrossRef]
39. Liang, Z.; Yi, L.; Huang, Z.; Lu, B.; Jiang, X.; Cai, W.; Tian, B.; Jin, Y. Insight of Iron Ore-Coal Composite Reduction in a Pilot Scale Rotary Kiln: A Post-Mortem Study. *Powder Technol.* **2019**, *356*, 691–701. [CrossRef]
40. Morales, R.G. Flexibility in Using Iron Ores for Direct Reduction. In Proceedings of the Metal Bulletin-2nd Latin American Steel & Iron Ore Conference, Rio de Janeiro, Brazil, April 2000; Available online: https://www.market-ing.com.mx/images/descargas/esp/MB_Paper.pdf (accessed on 27 July 2024).

41. Lu, Q.; Jiang, W.F.; Lu, C.X.; Zhao, L.G.; Yin, H.S.; Lang, J.F. Carbonising Mechanism and Carbon Distribution Behaviour during Direct Reduction in Shaft Furnace. *Ironmak. Steelmak.* **1999**, *26*, 122–126. [CrossRef]
42. Van den Berg, J.C.; Dippenaar, R.J. Fluidized-Bed Reduction of Fine Iron Ore by the in Situ Combustion of Coal. *J. S. Afr. Inst. Min. Metall.* **1989**, *89*, 89–98.
43. Komatina, M.; Gudenau, H.W. The Sticking Problem during Direct Reduction of Fine Iron Ore in the Fluidized Bed. *Metall. Mater. Eng.* **2004**, *10*, 309–328. [CrossRef]
44. Sohn, I.; Fruehan, R.J. The Reduction of Iron Oxides by Volatiles in a Rotary Hearth Furnace Process: Part I. The Role and Kinetics of Volatile Reduction. *Met. Mater. Trans. B* **2005**, *36*, 605–612. [CrossRef]
45. Nyakudya, R.Y.; Ayomoh, M. Sustainability Enhancement of the Coal Based Direct Reduction of Iron Premised on a Rotary Kiln. In *Manufacturing Driving Circular Economy*; Kohl, H., Seliger, G., Dietrich, F., Eds.; Lecture Notes in Mechanical Engineering; Springer International Publishing: Cham, Switzerland, 2023; pp. 211–218, ISBN 978-3-031-28838-8.
46. Zhu, D.; Mendes, V.; Chun, T.; Pan, J.; Li, Q.; Li, J.; Qiu, G. Direct Reduction Behaviors of Composite Binder Magnetite Pellets in Coal-Based Grate-Rotary Kiln Process. *ISIJ Int.* **2011**, *51*, 214–219. [CrossRef]
47. Wolfinger, T.; Spreitzer, D.; Schenk, J. Analysis of the Usability of Iron Ore Ultra-Fines for Hydrogen-Based Fluidized Bed Direct Reduction—A Review. *Materials* **2022**, *15*, 2687. [CrossRef]
48. Narita, K.; Kaneko, D.; Kimura, Y.; Takenaka, Y.; Tanaka, H.; Inada, Y. Production of Reduced Iron by Model Plant of Shaft Furnace. *Tetsu-to-Hagane* **1981**, *67*, 508–517. [CrossRef] [PubMed]
49. Lee, G.-Y.; Song, J.; Lee, J.-S. Reaction Kinetics and Phase Transformation during Hydrogen Reduction of Spherical Fe_2O_3 Nanopowder Agglomerates. *Powder Technol.* **2016**, *302*, 215–221. [CrossRef]
50. Efe Kinaci, M.; Lichtenegger, T.; Schneiderbauer, S. Direct Reduction of Iron-Ore in Fluidized Beds. In *Computer Aided Chemical Engineering*; Elsevier: Amsterdam, The Netherlands, 2018; Volume 43, pp. 217–222, ISBN 978-0-444-64235-6.
51. Battle, T.; Srivastava, U.; Kopfle, J.; Hunter, R.; McClelland, J. The Direct Reduction of Iron. In *Treatise on Process Metallurgy*; Elsevier: Amsterdam, The Netherlands, 2014; pp. 89–176, ISBN 978-0-08-096988-6.
52. Yang, Y.; Raipala, K.; Holappa, L. Ironmaking. In *Treatise on Process Metallurgy*; Elsevier: Amsterdam, The Netherlands, 2014; pp. 2–88, ISBN 978-0-08-096988-6.
53. Baig, S. Cost Effectiveness Analysis of HYL and Midrex DRI Technologies for the Iron and Steel-Making Industry. Master's Thesis, Duke University, Durham, NC, USA, 2016.
54. Nduagu, E.I.; Yadav, D.; Bhardwaj, N.; Elango, S.; Biswas, T.; Banerjee, R.; Rajagopalan, S. Comparative Life Cycle Assessment of Natural Gas and Coal-Based Directly Reduced Iron (DRI) Production: A Case Study for India. *J. Clean. Prod.* **2022**, *347*, 131196. [CrossRef]
55. Kumar, B.; Mishra, S.; Roy, G.G.; Sen, P.K. Estimation of Carbon Dioxide Emissions in Rotary Hearth Furnace Using a Thermodynamic Model. *Steel Res. Int.* **2017**, *88*, 1600265. [CrossRef]
56. Yang, Y.; Raipala, K.; Holappa, L. Future of Process Metallurgy. In *Treatise on Process Metallurgy*; Elsevier: Amsterdam, The Netherlands, 2014; pp. 1563–1726, ISBN 978-0-08-096988-6.
57. Ishikawa, H.; Kopfle, J.; McClelland, J.; Ripke, J. Rotary Hearth Furnace Technologies for Iron Ore and Recycling Applications. *Arch. Metall. Mater.* **2008**, *53*, 541–545.
58. Morris, A.E. Iron Resources and Direct Iron Production. In *Encyclopedia of Materials: Science and Technology*; Elsevier: Amsterdam, The Netherlands, 2001; pp. 4302–4310, ISBN 978-0-08-043152-9.
59. Soeparto, A.B. The Effect of Oxy-Carbon Injection on EAF Efficiency. Master's Thesis, University of Wollongong, Wollongong, Australia, 1996.
60. Kim, W.; Sohn, I. Critical Challenges Facing Low Carbon Steelmaking Technology Using Hydrogen Direct Reduced Iron. *Joule* **2022**, *6*, 2228–2232. [CrossRef]
61. Monsen, B.; Thomassen, E.; Bragstad, I.; Ringdalen, E.; Hoegaas, P. Characterization of DR Pellets for DRI Applications. In Proceedings of the AISTech 2015 Proceedings; Association for Iron and Steel Technology: Cleveland, OH, USA, 2015.
62. International Iron Metallics Association. Available online: https://www.metallics.org/ (accessed on 14 April 2024).
63. Small, M. Direct Reduction of Iron Ore. *JOM* **1981**, *33*, 67–75. [CrossRef]
64. Paknahad, P.; Askari, M.; Shahahmadi, S.A. Cold-Briquetted Iron and Carbon (CBIC), Investigation of Steelmaking Behavior. *J. Mater. Res. Technol.* **2020**, *9*, 6655–6664. [CrossRef]
65. Direct Reduced Iron (DRI). Available online: https://www.metallics.org/dri.html (accessed on 11 March 2024).
66. Abd Elkader, M.; Fathy, A.; Eissa, M.; Shama, S. Effect of Direct Reduced Iron Proportion in Metallic Charge on Technological Parameters of EAF Steelmaking Process. *Int. J. Sci. Res.* **2016**, *5*, 2016–2024.
67. Tappeiner, T. Ganzheitliche Betrachtung Des Einsatzes von LRI (Low Reduced Iron) Im Hochofen Zur CO_2-Minimierung. Ph.D. Thesis, Montanuniversität Leoben, Leoben, Austria, 2011.
68. Ahmed, H.; Sandeep Kumar, T.K.; Alatalo, J.; Björkman, B. Effect of Carbon Concentration and Carbon Bonding Type on the Melting Characteristics of Hydrogen-Reduced Iron Ore Pellets. *J. Mater. Res. Technol.* **2022**, *21*, 1760–1769. [CrossRef]
69. Calderon Hurtado, F.A.; Govro, J.; Emdadi, A.; O'Malley, R.J. The Melting Behavior of Hydrogen Direct Reduced Iron in Molten Steel and Slag: An Integrated Computational and Experimental Study. *Metals* **2024**, *14*, 821. [CrossRef]

70. Linklater, J. Adapting to Raw Materials Challenges: Part 1—Operating MIDREX Plants with Lower Grade Pellets & Lump Ores. Available online: https://www.midrex.com/tech-article/adapting-to-raw-materials-challenges-part-1-operating-midrex-plants-with-lower-grade-pellets-lump-ores/ (accessed on 7 March 2024).
71. Barati, M. Application of Slag Engineering Fundamentals to Continuous Steelmaking. In *Treatise on Process Metallurgy*; Elsevier: Amsterdam, The Netherlands, 2014; pp. 305–357, ISBN 978-0-08-096984-8.
72. Ma, Y.; Souza Filho, I.R.; Bai, Y.; Schenk, J.; Patisson, F.; Beck, A.; Van Bokhoven, J.A.; Willinger, M.G.; Li, K.; Xie, D.; et al. Hierarchical Nature of Hydrogen-Based Direct Reduction of Iron Oxides. *Scr. Mater.* **2022**, *213*, 114571. [CrossRef]
73. Gyllenram, R.; Sikstrom, P.; Hahne, R.; Tottie, M. *Classification of DRI/HBI Based on the Performance in the EAF A Help for Steelmaker's Procurement of Metallics*, METEC ESTAD: Düsseldorf, Germany,, 2015.
74. Tokuda, M.; Yoshikoshi, H.; Ohtani, M. Kinetics of the Reduction of Iron Ore. *ISIJ Int.* **1973**, *13*, 350–363. [CrossRef]
75. El-Geassy, A.A. Influence of Doping with CaO and/or MgO on Stepwise Reduction of Pure Hematite Compacts. *Ironmak. Steelmak.* **1999**, *26*, 41–52. [CrossRef]
76. Gonzalez, R.L.; Acosta, F.L.; Lowry, M.; Kundrat, D.; Wyatt, A.; Kuntze, J.; Fuchs, H. *Improvements in Yield in an All-DRI-Fed EAF from Minimization of FeO Generation during Melting as Well as Post-Reduction of FeO from Residual Slag*; Iron and Steel Technology: Nashville, TN, USA, 2018; pp. 36–41.
77. Manning, C.P.; Chevrier, V.F. Maximizing Iron Unit Yield from Ore to Liquid Steel (Part 2—DRI Physical Properties and DRI Handling and Storage). Available online: https://www.midrex.com/tech-article/maximizing-iron-unit-yield-from-ore-to-liquid-steel-part-2-dri-physical-properties-and-dri-handling-and-storage/ (accessed on 7 March 2024).
78. Sane, A.; Buragino, G.; Makwana, A.; He, X. Enhancing Direct Reduced Iron (DRI) for Use in Electric Steelmaking. Available online: https://www.airproducts.com/-/media/files/en/335/335-20-002-us-enhancing-direct-reduced-iron-43251.pdf (accessed on 7 March 2024).
79. Heo, J.H.; Park, J.H. Effect of Direct Reduced Iron (DRI) on Dephosphorization of Molten Steel by Electric Arc Furnace Slag. *Met. Mater. Trans. B* **2018**, *49*, 3381–3389. [CrossRef]
80. Hwang, H.-S.; Chung, U.-C.; Chung, W.-S.; Cho, Y.-R.; Jung, B.-H.; Martin, G.P. Carburization of Iron Using $CO-H_2$ Gas Mixture. *Met. Mater. Int.* **2004**, *10*, 77–82. [CrossRef]
81. He, Y.; Pistorius, P.C. Laboratory Carburization of Direct-Reduced Iron in CH_4-H_2-N_2 Gas Mixtures, and Comparison with Industrial Samples. *Met. Mater. Trans. B* **2016**, *47*, 1538–1541. [CrossRef]
82. Longbottom, R.J.; Ostrovski, O.; Zhang, J.; Young, D. Stability of Cementite Formed from Hematite and Titanomagnetite Ore. *Met. Mater. Trans. B* **2007**, *38*, 175–184. [CrossRef]
83. Kumar, T.K.S.; Alatalo, J.; Ahmed, H.; Björkman, B. Effect of Temperature and Gas Mixtures on Cementite Formation During the Carburization of Hydrogen-Reduced DRI. *J. Sustain. Metall.* **2022**, *8*, 1450–1464. [CrossRef]
84. Olsson, R.G.; Turkdogan, E.T. Catalitic Effect of Iron on Decomposition of Carbon Monoxide: II. Effect of Additions of H_2, H_2O, CO_2, SO_2 and H_2S. *Metall. Trans.* **1974**, *5*, 21–26. [CrossRef]
85. Fruehan, R.J. The Rate of Carburization of Iron in CO-H_2 Atmospheres: Part I. Effect of Temperature and CO and H_2-Pressures. *Met. Trans.* **1973**, *4*, 2123–2127. [CrossRef]
86. Iguchi, Y.; Endo, S. Carburized Carbon Content of Reduced Iron and Direct Carburization in Carbon Composite Iron Ore Pellets Heated at Elevated Temperature. *ISIJ Int.* **2004**, *44*, 1991–1998. [CrossRef]
87. Grabke, H.J.; Müller-Lorenz, E.M.; Schneider, A. Carburization and Metal Dusting on Iron. *ISIJ Int.* **2001**, *41*, S1–S8. [CrossRef] [PubMed]
88. Hayashi, S.; Iguchi, Y. Production of Iron Carbide from Iron Ores in a Fluidized Bed. *ISIJ Int.* **1998**, *38*, 1053–1061. [CrossRef]
89. Kim, G. Carbon Concentration and the Use of Direct-Reduced Iron in Ironmaking and Steelmaking. Ph.D. Thesis, Carnegie Mellon University, Pittsburgh, PA, USA, 2020.
90. Kim, G.; Kacar, Y.; Pistorius, P.C. Carbon Bonding State Has a Small Effect on Melting of Direct-Reduced Iron. *Met. Mater. Trans. B* **2019**, *50*, 2508–2516. [CrossRef]
91. Jess, A.; Grabke, H.; Steffen, R. *Reoxidation and Ignition Behaviour of DRI to Improve Safety*; European Commission, Directorate-General for Research and Innovation: Luxemburg, 2003.
92. Majhi, T.R. Modeling of Rotary Kiln for Sponge Iron Processing. Master's Thesis, National Institute of Technology, Rourkela, India, 2012.
93. Pietsch, N. The Influence of Raw Material and Reduction Temperature on the Structure and Characteristics of DRI. *SME-AIME* **1978**, *264*, 1784.
94. AbdElmomen, S.S. Reoxidation of Direct Reduced Iron in Ambient Air. *Ironmak. Steelmak.* **2014**, *41*, 107–111. [CrossRef]
95. Cavaliere, P.; Dijon, L.; Laska, A.; Koszelow, D. Hydrogen Direct Reduction and Reoxidation Behaviour of High-Grade Pellets. *Int. J. Hydrog. Energy* **2024**, *49*, 1235–1254. [CrossRef]
96. Paknahad, P.; Askari, M. Modeling, kinetics investigation and determining the controlling mechanisms of atmospheric oxidation of "cold" briquetted iron and carbon (CBIC). *Met. Eng.* **2021**, *23*, 206–219. [CrossRef]
97. Yazir, D.; Sahin, B.; Alkac, M. Selection of an Inert Gas System for the Transportation of Direct Reduced Iron. *Math. Probl. Eng.* **2021**, *2021*, 8529724. [CrossRef]
98. Towhidi, N. Influence of Direct Reduction Condition of Hematite Pellets with H_2/CO on the Oxidation Behaviour of DRI in Air. *Steel Res. Int.* **2003**, *74*, 595–600. [CrossRef]

99. El-Geassy, A.A.; El-Kashif, F.O.; Nasr, M.I.; Omar, A.A. Kinetics and Mechanisms of Re-Oxiation of Freshly Reduced Iron Compacts. *ISIJ Int.* **1994**, *34*, 541–547. [CrossRef]
100. Abd Elmomen, S.S. Reoxidation of Direct Reduced Iron in Stagnant Air in The Temperature Range between 150 and 450°C. *Bull. Tabbin Inst. Metall. Stud. (TIMS)* **2021**, *109*, 72–86. [CrossRef]
101. Upadhya, K. The Kinetics and Mechanism(s) of Oxidation of Direct Reduced Iron. *JOM* **1984**, *36*, 39–43. [CrossRef]
102. Medina, J.O. Studies of the Re-Oxidation Behavior of DRI in Air at Moderate Temperatures. Master's Thesis, Michigan Tech. University, Houghton, MI, USA, 1981.
103. Bandopadhyay, A.; Ganguly, A.; Gupta, K.N.; Ray, H.S. Investigations on the Anomalous Oxidation Behaviour of High-Carbon Gas-Based Direct Reduced Iron (DRI). *Thermochim. Acta* **1996**, *276*, 199–207. [CrossRef]
104. Bandopadhyay, A.; Ganguly, A.; Prasad, K.K.; Sarkar, S.B.; Ray, H.S. Determination of Kinetic Parameters for the Reoxidation of Direct Reduced Iron under Rising Temperature Conditions. *Thermochim. Acta* **1993**, *228*, 271–281. [CrossRef]
105. Hunter, R.L. *Direct Reduced Iron: Technology and Economics of Production and Use*; ISS-AIME: Warrendale, PA, USA, 1999.
106. Merki, A.; Rothberger, J.; Millner, R.; Sterrer, W. The New Age of HBI. Available online: https://magazine.primetals.com/2023/02/28/the-new-age-of-hbi/ (accessed on 11 March 2024).
107. Understanding the Different Direct Reduced Iron Products. Available online: https://gard.no/articles/understanding-the-different-direct-reduced-iron-products/ (accessed on 27 July 2024).
108. Hot Briquetted Iron (HBI). Available online: https://www.metallics.org/hbi.html (accessed on 15 April 2024).
109. Daghagheleh, O.; Schenk, J.; Zheng, H.; Taferner, B.; Forstner, A.; Rosenfellner, G. Long-Term Reoxidation of Hot Briquetted Iron in Various Prepared Climatic Conditions. *Steel Res. Int.* **2023**, *94*, 2200535. [CrossRef]
110. Nagel, H. Die Oxydation Des Eisenschwamms, Die Fakultät Für Bergbau Und Hüttenwesen Der Rheinisch. Ph.D. Thesis, Technical University of Aachen, Aachen, Germany, 1973.
111. Das, S.; Hendry, M.J. Application of Raman Spectroscopy to Identify Iron Minerals Commonly Found in Mine Wastes. *Chem. Geol.* **2011**, *290*, 101–108. [CrossRef]
112. Barrington, C. Safe Shipment of HBI—HBI C-FLEX Project & IMO Regulatory Update. Available online: https://www.midrex.com/tech-article/safe-shipment-of-hbi-hbi-c-flex-project-imo-regulatory-update/ (accessed on 2 April 2024).
113. Barrington, C. Hot Briquetted Iron-c-Flex Project: Addressing a Challenge to the HBI Value Chain. Available online: https://www.midrex.com/tech-article/hot-briquetted-iron-c-flex-project-addressing-a-challenge-to-the-hbi-value-chain/ (accessed on 3 April 2024).
114. Gray, J.; Saha-Chaudhury, N.; Sahajwalla, V. Characterisation and Corrosion of Laboratory Scale Briquettes of Reduced Iron. *ISIJ Int.* **2002**, *42*, 826–833. [CrossRef]
115. HBI C-Flex. Available online: https://hbi-c-flex.eu/ (accessed on 4 April 2024).
116. DRI Products and Applications. Providing Flexibility for Steelmaking. Available online: https://www.midrex.com/wp-content/uploads/MIdrexDRI_ProductsBrochure_4-12-18-1.pdf (accessed on 23 April 2024).
117. Ravenscroft, C.; Hunter, R.; Griscom, F. The Versatile OBM (Ore-Based Metallic): Part 2—How to Get What You Paid for: A Guide to Maintaining the Value of DRI. Available online: https://www.midrex.com/tech-article/the-versatile-obm-ore-based-metallic-part-2-how-to-get-what-you-paid-for-a-guide-to-maintaining-the-value-of-dri/ (accessed on 12 March 2024).
118. Ahmad, J.K. Inhibition of Reoxidation of Direct Reduced Iron (DRI) or sponge iron. *Int. J. Mater. Sci. Appl.* **2015**, *4*, 7–10. [CrossRef]
119. Ahmed, K.A. Inhibition of Reoxidation of Direct Reduced Iron. *ISIJ Int.* **1984**, *24*, 163–164. [CrossRef]
120. Iwamoto, I.; Kurniawan, A.; Hasegawa, H.; Kashiwaya, Y.; Nomura, T.; Akiyama, T. Reduction Behaviors and Generated Phases of Iron Ores Using Ammonia as Reducing Agent. *ISIJ Int.* **2022**, *62*, 2483–2490. [CrossRef]
121. Ma, Y.; Bae, J.W.; Kim, S.; Jovičević-Klug, M.; Li, K.; Vogel, D.; Ponge, D.; Rohwerder, M.; Gault, B.; Raabe, D. Reducing Iron Oxide with Ammonia: A Sustainable Path to Green Steel. *Adv. Sci.* **2023**, *10*, 2300111. [CrossRef] [PubMed]
122. McCafferty, E. Effect of Ion Implantation on the Corrosion Behavior of Iron, Stainless Steels, and Aluminum—A Review. *Corrosion* **2001**, *57*, 1011–1029. [CrossRef]
123. Hosokai, S.; Kasiwaya, Y.; Matsui, K.; Okinaka, N.; Akiyama, T. Ironmaking with Ammonia at Low Temperature. *Environ. Sci. Technol.* **2011**, *45*, 821–826. [CrossRef]
124. Yasuda, N.; Mochizuki, Y.; Tsubouchi, N.; Akiyama, T. Reduction and Nitriding Behavior of Hematite with Ammonia. *ISIJ Int.* **2015**, *55*, 736–741. [CrossRef]
125. Durnovich, D.; Miller, T. The Basics of DRI Plant Safety. Available online: https://www.proquest.com/openview/69da5c8a7bbaf3345542e09e2cf683de/1.pdf?pq-origsite=gscholar&cbl=1056347#:~:text=Any%20DRI%20at%20a%20temperature,more%20than%20one%20metre%20high.&text=Bins%20and%20silos%20should%20be,inert%20gas%20from%20the%20bottom.&text=Top%20slide%20gates%20should%20be%20closed%20except%20when%20DRI%20is%20being%20delivered (accessed on 11 March 2024).
126. Djadjev, I. The Evolving Law and Regulation of the Carriage of Dangerous Goods by Sea The IMDG Code and the IMSBC Code. *SSRN J.* **2015**, 1–8. [CrossRef]
127. Dutta, S.K.; Sah, R. Worldwide Direct Reduced Iron Scenario and Hazards Associated with Their Storage and Shipments. *Iron Steel Rev.* **2014**, *58*, 148–156.

128. Direct Reduced Iron (DRI). Available online: https://www.cargohandbook.com/Direct_Reduced_Iron_(DRI) (accessed on 11 March 2024).
129. Durinck, D.; Gurlit, W.; Muller, F.; van Albada, B. Closing Europe's Green-Metallics Gap. Available online: https://www.mckinsey.com/industries/metals-and-mining/our-insights/closing-europes-green-metallics-gap (accessed on 9 March 2024).
130. Sanjal, S. The Value of DRI—Using the Product for Optimum Steelmaking. Available online: https://www.midrex.com/tech-article/the-value-of-dri-using-the-product-for-optimum-steelmaking/#:~:text=versus%20Charge%20Material-,USING%20DRI,),%20and%20less%20back%20charging (accessed on 9 March 2024).
131. Voelker, B. Getting the Most from Direct Reduced Iron—Operational Results of MIDREX® Hot Transport-Hot Charging. Available online: https://www.midrex.com/tech-article/getting-the-most-from-direct-reduced-iron-operational-results-of-midrex-hot-transport-hot-charging/ (accessed on 9 March 2024).
132. Li, J.; Barati, M. Kinetics and Mechanism of Decarburization and Melting of Direct-Reduced Iron Pellets in Slag. *Met. Mater. Trans. B* **2009**, *40*, 17–24. [CrossRef]
133. Millner, R.; Rothberger, J.; Rammer, B.; Boehm, C.; Sterrer, W.; Hanspeter, O.; Chevrier, V. MIDREX H2—The Road to CO2-Free Direct Reduction. Available online: https://www.primetals.com/fileadmin/user_upload/landing_pages/2021/Green_Steel/Publications/downloads/AISTech_2021_MIDREX_H2_Final.pdf (accessed on 9 March 2024).
134. González, O.J.P.; Ramírez-Argáez, M.A.; Conejo, A.N. Mathematical Modeling of the Melting Rate of Metallic Particles in the Electric Arc Furnace. *ISIJ Int.* **2010**, *50*, 9–16. [CrossRef]
135. Gyllenram, R.; Arzpeyma, N.; Wei, W.; Jönsson, P.G. Driving Investments in Ore Beneficiation and Scrap Upgrading to Meet an Increased Demand from the Direct Reduction-EAF Route. *Min. Econ.* **2022**, *35*, 203–220. [CrossRef]
136. Pfeiffer, A.; Ernst, D.; Zheng, H.; Wimmer, G.; Schenk, J. The Behavior of Direct Reduced Iron in the Electric Arc Furnace Hotspot. *Metals* **2023**, *13*, 978. [CrossRef]
137. Morales, R.D.; Rubén, L.G.; López, F.; Camacho, J.; Romero, J.A. The Slag Foaming Practice in EAF and Its Influence on the Steelmaking Shop Productivity. *ISIJ Int.* **1995**, *35*, 1054–1062. [CrossRef]
138. Birley, R.I. *Decarbonising the DRI Feed for EAF Using H_2*; Materials Processing Institute: Sheffield, UK, 2021.
139. Turcotte, S.; Marquis, H.; Dancy, T. The Use of Direct Reduced Iron in the Electric Arc Furnace. In *Electric Furnace Steelmaking*; AIME, Iron Steel Society: Warrendale, PA, USA, 1985; pp. 115–126.
140. Kieush, L.; Schenk, J.; Koveria, A.; Hrubiak, A.; Hopfinger, H.; Zheng, H. Evaluation of Slag Foaming Behavior Using Renewable Carbon Sources in Electric Arc Furnace-Based Steel Production. *Energies* **2023**, *16*, 4673. [CrossRef]
141. Kieush, L.; Schenk, J. Influence of Carbon Material Properties on Slag-Foaming Dynamics in Electric Arc Furnaces: A Review. *Steel Res. Int.* **2024**, 2400235. [CrossRef]
142. Kieush, L. Coal Pyrolysis Products Utilisation for Synthesis of Carbon Nanotubes. *Pet. Coal* **2019**, *61*, 461–463.
143. Kurecki, M.; Meena, N.; Shyrokykh, T.; Korobeinikov, Y.; Jarnerud Örell, T.; Voss, Z.; Pretorius, E.; Jones, J.; Sridhar, S. Recycling Perspectives of Electric Arc Furnace Slag in the United States: A Review. *Steel Res. Int.* **2024**, 2300854. [CrossRef]
144. Memoli, F. Behavior and Benefits of High Fe_3C-DRI in the EAF. *Iron Steel Technol.* **2019**, *2*, 1928–1945.
145. Lule, R.; Lopez, F.; Espinoza, J.; Torres, R.; Molares, R.D. The Production of Steels Applying 100% DRI for Nitrogen Removal, the Experience of ArcelorMittal Lazaro Cardenas Flat Carbon. In Proceedings of the AISTech 2009—Iron and Steel Technology Conference, St. Louis, MO, USA, 4 May 2009; Volume 1, pp. 489–498.
146. Dutta, S.K.; Lele, A.B.; Pancholi, N.K. Studies on Direct Reduced Iron Melting in Induction Furnace. *Trans. Indian Inst. Met.* **2004**, *57*, 467–473.
147. Hassan, A.; Kotelnikov, G.; Abdelwahed, H. Melting Characteristics of Alternative Charging Materials in an Electric Arc Furnace Steelmaking. *Ironmak. Steelmak.* **2021**, *48*, 1136–1141. [CrossRef]
148. Song, S.; Zhao, J.; Pistorius, P.C. MgO Refractory Attack by Transient Non-Saturated EAF Slag. *Met. Mater. Trans. B* **2020**, *51*, 891–897. [CrossRef]
149. Heo, J.; Park, J.H. Interfacial Reactions between Magnesia Refractory and Electric Arc Furnace (EAF) Slag with Use of Direct Reduced Iron (DRI) as Raw Material. *Ceram. Int.* **2022**, *48*, 4526–4538. [CrossRef]
150. Heo, J.; Park, J.H. Effect of Temperature on the Slag/Refractory Interfacial Reaction with Directed Reduced Iron (DRI) Addition in an Electric Arc Furnace (EAF) Process: Diffusional Growth of Magnesiowüstite Layer by Boltzmann-Matano Analysis. *Ceram. Int.* **2022**, *48*, 17217–17224. [CrossRef]
151. Mulholland, E.W.; Hazeldean, G.S.F.; Davies, M.W. Visualization of Slag-Metal Reactions by X-RAY Fluoroscopy: Decarburization in Basic Oxygen Steelmaking. *J. Iron Steel Inst.* **1973**, *211*, 632–639.
152. Min, D.-J.; Fruehan, R.J. Rate of Reduction of FeO in Slag by Fe-C Drops. *Met. Trans. B* **1992**, *23*, 29–37. [CrossRef]
153. Goldstein, D.A.; Fruehan, R.J.; Oztruk, B. The Behaviour of DRI in Slag-Metal Systems and Its Effect on the Nitrogen Content of Steel. *Iron Steelmak.* **1999**, *26*, 49–61.
154. Sharifi, E.; Barati, M. The Reaction Behavior of Direct Reduced Iron (DRI) in Steelmaking Slags: Effect of DRI Carbon and Preheating Temperature. *Met. Mater. Trans. B* **2010**, *41*, 1018–1024. [CrossRef]
155. Kiasaraei, E.S. Decarburization and Melting Behavior of Direct-Reduced Iron Pellets in Steelmaking Slag. Master's Thesis, University of Toronto, Toronto, ON, Canada, 2010.

156. Murthy, G.G.K.; Sawada, Y.; Elliott, J.F. Reduction of FeO Dissolved in CaO-SiO_2-Al_2O_3 Slags by Fe—C Droplets. *Ironmak. Steelmak.* **1993**, *20*, 179–190.
157. Sadrnezhaad, K. Direct Reduced Iron: An Advantageous Charge Material for Induction Furnaces. *J. Eng.* **1990**, *3*, 37–47.

MDPI AG
Grosspeteranlage 5
4052 Basel
Switzerland
Tel.: +41 61 683 77 34

Metals Editorial Office
E-mail: metals@mdpi.com
www.mdpi.com/journal/metals

www.ingramcontent.com/pod-product-compliance
Lightning Source LLC
Chambersburg PA
CBHW061618210926
51850CB00009B/2364
9783725862306